Chemistry in Your Kitchen

Chemistry in Your Kitchen

Matthew Hartings
American University, Washington, DC, USA
Email: hartings@american.edu

THE QUEEN'S AWARDS
FOR ENTERPRISE:
INTERNATIONAL TRADE
2013

Print ISBN: 978-1-78262-313-7

A catalogue record for this book is available from the British Library

© Matthew Hartings 2017

All rights reserved

Apart from fair dealing for the purposes of research for non-commercial purposes or for private study, criticism or review, as permitted under the Copyright, Designs and Patents Act 1988 and the Copyright and Related Rights Regulations 2003, this publication may not be reproduced, stored or transmitted, in any form or by any means, without the prior permission in writing of The Royal Society of Chemistry or the copyright owner, or in the case of reproduction in accordance with the terms of licences issued by the Copyright Licensing Agency in the UK, or in accordance with the terms of the licences issued by the appropriate Reproduction Rights Organization outside the UK. Enquiries concerning reproduction outside the terms stated here should be sent to The Royal Society of Chemistry at the address printed on this page.

The RSC is not responsible for individual opinions expressed in this work.

The authors have sought to locate owners of all reproduced material not in their own possession and trust that no copyrights have been inadvertently infringed.

Published by The Royal Society of Chemistry,
Thomas Graham House, Science Park, Milton Road,
Cambridge CB4 0WF, UK

Registered Charity Number 207890

Visit our website at www.rsc.org/books

Printed in the United Kingdom by CPI Group (UK) Ltd, Croydon, CR0 4YY, UK

To Erika who holds our whole family together.
Everything tastes better because of you.

Contents

Breakfast

Chapter 1 Coffee — 3

Flavor Creation — 6
Flavor Extraction — 15
References — 21

Chapter 2 Bacon — 23

Some Pig — 26
Pigs are Magic — 29
Maillard is Magic Too — 30
Food Safety and Evolution — 36
My Five Seconds of Fame — 38
References — 40

Chapter 3 Eggs — 42

Treat Eggs with the Respect they Deserve — 43
Incredible, Edible Science — 45
Egg-Cellent Research — 49
Chef Ramsay *vs.* My Mom — 54
References — 58

Chapter 4 Pancakes — 60

References — 77

Lunch

Chapter 5 Jelly — 81

- Let's Jam — 87
- Molecular Schmolecular — 94
- Just Jamming — 98
- References — 99

Chapter 6 Macaroni and Cheese — 100

- Milky Musings — 102
- On and Off — 107
- I Melt with You — 111
- References — 117

Chapter 7 Bread — 118

- Citizen Scientists — 124
- References — 134

Chapter 8 Vinaigrette — 135

- Back to the Basics — 152
- References — 153

Dinner

Chapter 9 Pizza — 157

- Water In. Water Out. Balance — 158
- Crust — 163
- Saucy — 167
- Say Cheese — 170
- Pizza Time — 171
- Delicious by Any Definition — 173
- References — 174

Chapter 10 Meat Time — 176

- It's About Time — 195
- References — 195

Contents ix

Chapter 11 More Meat Time	196
References	212

Chapter 12 Color	213
Jillian Dempsey	221
Morgan Cable	224
Gretchen Keller	227
Me	228
References	231

Drinks and Dessert

Chapter 13 Beer	235
Brewing: Biology and Chemistry as Art	237
All About that Base	240
A Monster Mash	245
Why So Bitter?	250
Sending Them Off	252
References	253

Chapter 14 Cocktails	254
References	273

Chapter 15 Ice Cream	274
Freezing	276
Churning	283
The Ice Cream Base	287
Making Not Ice Cream	288
References	290

Chapter 16 Pie	292
Extreme 1: Bread	297
Extreme 2: Phyllo	298
Pie	301
Epilogue	305
References	305

Subject Index	**307**

Breakfast

CHAPTER 1

Coffee

We start with a humble cup of coffee.

A simple act that is repeated around the world over 2 billion times a day,[1] brewing and drinking a cup of coffee can seem to many as reflexive as blinking, yawning, or taking a look at their phones in the middle of a meeting.

And yet, to get from the farm to your face, a coffee bean must become a player in a well-conducted symphony of chemistry.

This description does not aim to highlight the single-sourced, fair trade, small-batch roasted, hand ground, 30 g of beans, and 300 mL of 88 °C water Pour-Over coffee produced by the barista who knows more about coffee than any human really should. No, every little bit of coffee – from the most lackluster, office breakroom swill to that over-crafted Pour-Over – is chemistry in a cup.

As a chemist, I am continuously amazed and inspired by the intricate chemical processes that are required to complete even the most unpretentious recipe. I am in awe of every cook who, when in the kitchen, masters an interconnected network of competing chemical reactions. These same reactions, when written and diagramed, would send even the most grizzled laboratory veteran running for the hills.

Because of this, I see cooking as a vehicle for having a conversation about chemistry with people who would normally rather

---

Chemistry in Your Kitchen
By Matthew Hartings
© Matthew Hartings, 2017
Published by the Royal Society of Chemistry, www.rsc.org

not think about it. The sheer number of people who are willing to seek out that singular barista or who meticulously experiment with their own recipes are testament to this. In fact, of all of the things I have written, the piece that has received the most attention is a blog post I wrote on the science of gin and tonic.[2] I think that it struck such a nerve because there are a lot of people out there who are passionate about gin, tonic, and gin and tonic. And, because of this passion, I was able to pose a really intriguing question: Why do gins and tonics taste so different on their own, but are inherently 'gin and tonic-y' when mixed together in a drink? What is the chemistry that controls this experience? I'm not entirely sure that I adequately answered that question in my blog post, (you can be sure that I will revisit gin and tonic later in this book); but, using a simple cocktail, I was able to discuss van der Waals interactions, protein-ligand binding thermodynamics, and neurochemistry with people who would normally be completely turned off by such topics. As the Pulitzer Prize-winning journalist, Deborah Blum, noted in a previous interview with me, "[Chemistry is not just] the story of some weird experiment done in a distant lab. It's the story of dinner, or something equally ordinary and therefore important."[3] I must admit that I agree with her sentiments. I find that there is an incredible amount of beauty and elegance in the chemistry of the common.

Another reason why food science topics resonate is the common ground between chemists and non-chemists. Specifically, no one really has a full understanding of what is going on. We (scientists) may have a pretty good clue about things. But, often, there are aspects of cooking chemistry, and cooking science, that elude us. Contrary to what some may believe, the absence of knowledge is very exciting to a scientist. And, as far as kitchen chemistry goes, in the absence of knowledge, science must turn to the people who practice and observe this type of chemistry on a daily basis: professional and home cooks.

In these situations, that elude scientific understanding, the practitioners (roasters, bakers, chefs, mixologists, brewers, and home cooks) are the experts. Scientists do well by listening to them, as I have hopefully done in this book. These people (consciously or not) know the important variables in their recipes. They understand that a specific change in color or texture or aroma can often mean that a recipe is ready to progress to its next step. Their observations are attuned to the food that they are cooking. In this respect, everyone is a chemist when they step into a kitchen. Everyone does chemistry when they cook. And, anyone is capable of making novel discoveries or observations when they are standing in front of an oven. I, for one, have learned a great deal about chemistry by talking with students, friends, relatives, and professional chefs alike. As Chef Gusteau says in *Ratatouille*, "Anyone can cook."[4] The same is true of chemistry. Anyone can be a chemist in the kitchen. For professional food scientists and professional chemists listening to these practitioners can be a very valuable experience.

From another point of view, I believe that cooking can be a great inspiration to chemists. Research chemistry is constantly moving from clean experiments to systems that are convoluted and complex and messier than their predecessors. Cooking is complex chemistry. Cooking can help to illuminate problems that I am having in my own lab. I see food as a lens through which I can learn how to be more creative in research. The cooking that humans have been doing for millennia can give clues towards the advancement of chemical science. While it is true that science and cooking have always been connected (the bain-marie, or double boiler, is thought to have been developed by an alchemist named Mary the Jewess[5]), I believe that kitchen chemistry can be a strong inspiration for modern day research.

This book is an exploration of these topics: discussing and sharing the really interesting chemistry that happens in our food because it's fun and so that we can be better cooks, looking for explanations of what goes on when we cook our food, and exploring the ways that food can teach us about chemical research.

And that brings us back to our cup of coffee. It turns out that this unassuming drink can tell us a lot about the most important processes in making food, it can illustrate how chemistry works, and by understanding these things, it can also show us how we can be better cooks and chemists.

If there were any food in this world where humans might expect to produce consistent "perfection", coffee might be it. The 2 billion cups that we consume every day is no joke. The practice and repetition that go into all of those vats of java is an undertaking of epic proportion. And yet, it is still exceedingly common to find coffee that doesn't match up with even our most mediocre expectations.

The quest for perfection also often falls short in coffee shops that are dedicated to finding, and serving, excellence. In Nathan Myhrvold's tome, *Modernist Cuisine*, he and his co-authors describe the elusive "god shot" of espresso.[6] This espresso is produced with exquisitely roasted beans, in a room at the ideal humidity, with the water at a perfect temperature and pressure and containing the right concentration of magnesium and calcium and carbonate ions,[7] in the second phase of the moon, while standing on one foot, with your eyes crossed. In the book, Myhrvold laments the infrequency of pulling off the perfect pull. ("Pull" is a term for making espresso that originates from a time when espresso machines required a lever to generate the pressure needed to force water through the coffee grounds.) There are some baristas and coffee pros who refute the notion of a "god shot."[8] They claim that most espressos are made well when the baristas are paying a little bit of attention. These professionals think that a "god shot" is more a reflection of the drinker's mood than anything else. There may be something to this claim, as we will discuss later in this book. But the fact remains that most coffee producers still strive for consistent perfection.

In truth, coffee (the best and the worst) is the result of both chance and controlled chemistry. And, while perfection is at the whim of personal preference, good and really good can be made by paying attention to flavor creation, flavor extraction, and presentation. Put another way: How do coffee berries turn into coffee beans? How do we get the flavor out of the beans and into water? And, how are we most happy when drinking coffee? The first two questions are inherently chemical and, by exploring them in the context of coffee, we can understand the basis for the chemistry of a lot of other food preparation. The final question involves more cultural and neurochemical cues than anything else. However, if we understand what we like (our personal preferences) from the outset, we will better be able to produce that during any type of food preparation.

FLAVOR CREATION

Compass Coffee is a relatively new shop in one DC neighborhood that is being gentrified. The storefront is sitting right on the edge of what is new, and what has suffered years of public neglect. All social commentary aside, Compass is not (yet) in a spot that is ideal for their business. When I went to Compass to talk about roasting coffee beans, it was 10 am on a workday. The line for ordering coffee extended out of the door. As I was early for our meeting, I got in line like everyone else. I struck up a conversation with a gentleman in front of me who was wearing cycling gear. He said that he had never ridden through this neighborhood before and had never heard of this coffee shop. But, when he saw the line, he couldn't not go in. It's pretty obvious that the owners were doing something that has struck a nerve.

It was no mistake that I went to Compass for this conversation about coffee and science. Michael Haft and Harrison Suarez, Compass's owners, had been profiled by the *Washington Post* and other news outlets for their use of research based methods and analytical data keeping for optimizing their product.[9] I had also previously interviewed them for *Chemical and Engineering News* with my journalist friend, Jessica Morrison, for a story on nitrogen infused cold brew.[10] So, I knew that they were up to date on all of their science homework.

Suarez and Haft met in the Marines where they started to share their love of coffee. Neither of them had a strong background of science training. But both of them wanted to do "right" by their coffee. So, they started reading and playing around and learning and experimenting, and learning more. They tell me that, while they may be more analytically minded than some other roasters, that doesn't mean their process lacks art or any room for taste or personal preference. Michael told me that many see roasting as some mystical process. But science, as they see it, just shows how you can best find your way into these predilections.

When they opened their shop and started putting their team together, they hired Brandon Warner as their head roaster. Suarez said of Warner, "With coffee roasting being as much an art as it is a science, we wanted someone on our team who would come to the process with an open mind. They had to be willing to experiment and to follow what their results showed them."

And, from what I know about Warner, that describes him pretty well. Warner was an undergraduate psychology major who came to DC to work as a writer. He keeps a trimmed auburn beard and his accent immediately gives him away as a Midwesterner, much like me. Warner said that he had originally gotten into both home beer brewing and home coffee roasting. His small apartment pretty much prevented him from being able to do anything interesting with respect to beer making, and he was shoehorned into roasting coffee, which he did with a cast iron pan. As he played around and got a little better, his techniques and equipment got a little more sophisticated. And now, as head roaster at Compass, he roasts 30 kg of beans at a time.

As we started chatting, I quickly professed an interest in the green coffee beans that they use. Warner hauled over a tub of beans from Guatemala. The tub is about the shape of a trash can, only about half the height – think of a fat R2-D2. Warner pulled the lid off and told me to stick my head in and take a whiff. The smell reminds me, overwhelmingly, of hay, and I immediately think of the freshly cut hay that I would use as a child to feed cows on my family's small farm This is a very stark childhood memory for me. And, I can't quite tell if there are any other strong aromas, or if my subconscious is preventing me from smelling anything else. But, honestly, this is kind of what I expected the beans to smell like. I don't know why. Perhaps it is because of their grassy color and that they look like overgrown hayseeds Next, he pulled out a tub of beans from Ethiopia. I lean over expecting something similar but am completely floored by the difference in smell. These beans have a funkier aroma. They may be funky because I don't recognize the smell right away. But they smell fruitier and sweeter than the Guatemalan beans. Brandon says that this difference is fairly generalizable between Central and South American beans as compared to beans from Africa.

Coffee is prepared from the fruit of the *Coffea* plant, which naturally grew in some parts of Africa and southern Asia. It seems as though the caffeine, and our addiction to it, are written into the DNA of *Coffea*.[11,12] Caffeine is an alkaloid molecule and can be toxic at high enough concentrations. Through natural selection, *Coffea* plants produce high amounts of caffeine in their leaves. At these increased levels, caffeine will discourage insects from eating coffee leaves. Furthermore, when the leaves fall to the ground and eventually become part of the soil, the high caffeine content in the dirt can prevent other plants from growing near the *Coffea* tree. And it turns out that *Coffea* evolution is even sneakier than just making sure that its trees aren't killed off. Caffeine gets used to ensure successful reproduction as well. Trees that produced caffeine in their flowers and, subsequently, their berries, were more likely to be pollinated. Bees and other pollinators like the hit of caffeine just as much as we do.

For all animals, from insects to mammals, alkaloids are poisonous at high doses. We recognize these molecules as a bitter taste in our mouths. However, at lower doses, alkaloids stimulate positive responses from animals and become addictive. Caffeine is only one example of this. Nicotine, morphine, and cocaine also fall into this class of chemicals. Through evolutionary trial and error, coffee trees found a way to dose their flowers with caffeine so that the pollinating insects that visited these trees would get a little hit of caffeine with their nectar and would want to come back for more. It seems as though our addiction to caffeine is inextricably linked to *Coffea* killing off its plant competition and warding off harmful insects.

While caffeine may be the reason we continue to go back to coffee (and tea, for that matter), there are other chemistries inherent to the plant and its harvest that are tied to its initial pull on us: the aroma. It's true that the aroma, as we recognize it in coffee comes from the roasting process. The green coffee beans whose smells surrounded me in those tubs at Compass were grassy and fruity. They were vegetal. Roasted coffee is warm and inviting and savory. And, while it's true that roasting is key, those enticing aromas don't come from nowhere. They are the fragments and re-juxtapositions of molecules that are present in the green coffee.

Naturally, then, what those molecules are, and their origins, are critical to what the bean will become. There are some differences between Arabica (*Coffea arabica*) and Robusta (*Coffea canephora*), some important ones being that Arabica has more fats and sugars and Robusta has more caffeine. And, certainly some differences come from *terrior*. That saying, "you are what you eat," is just as valid for plants as it is for us humans. The mineral and nutrient content of the soil where the coffee tree grows results in differences that we can absolutely detect. But what interests me more is another aspect of *terrior* that shows up in aroma development.

Before I get to that, though, a brief description of coffee berry harvesting and pre-roasting preparation is necessary. A ripe coffee cherry is reminiscent of a cranberry, vibrant red and oblong.

After large scale harvesting, the ripe cherries are separated from the unripe and, consequently, pulped so that the seed (or bean) is removed of its flesh. Once the bean is isolated, one step remains in preparing it before roasting. Each seed features a protective film, called mucilage, which is a complex mixture of proteins and carbohydrates. While the berry is whole, the mucilage is a last line of protection for the seed against any natural toxins that might eventually destroy it. This film also prevents consistent quality roasting. So, it must be removed. It turns out that the historical (and industrially approved) way of removing the film is to let bacteria and yeast chew it off, *i.e.* fermentation.[13] There are two ways of doing this. One way involves "wet" fermentations in an industrial setting and lasts between 1 and 2 days. Another way ("dry" fermentation), involves spreading the seeds out and letting natural yeast and bacteria do their work over about 2 weeks. Even though wet fermentation reduces the risk of damage from insects, most beans undergo dry fermentation because of its low cost. Beans that are fermented industrially are usually from coffee plants that are grown in climates that are, in general, less hospitable for coffee production. This coffee tends to be the Robusta variety.

While both methods will remove the mucilage from the seed, there are important differences in the chemical outcomes of industrial and natural fermentation. The mucilage itself is made, mostly, from proteins and pectin, a polysaccharide like cornstarch or cellulose. We use pectin to make jellies and jams. To break down the mucilage, the microbes (bacteria and yeast) will chop pectin into smaller chemical pieces. The identity of these pieces will define the character of the roasted coffee. Dry fermentation tends to produce more simple sugars (like fructose and glucose, which are commonly known, and arabinose and galactose, which are likely new names to most readers). Wet fermentation produces more glutamic and aspartic acids, which result from protein breakdown as opposed to the sugars that result from pectin breakdown seen in dry fermentation. These simple sugars are going to be critical for the roasting process.

Another reason why the dry method may produce better beans than the wet process, is time. Microbes use specific proteins, called pectinases, to turn pectin from the long chains that give stability to the mucilage, into simple sugars. As pectinases do their work, there are other microbial proteins that can make chemicals that enhance the flavor and aroma of coffee. One issue with allowing these other proteins to do their thing is that they require much more time than pectinases do. That is, in the time it takes pectinase to chew off a single glucose unit from pectin, these other aroma-making proteins are only a fraction of the way through their work. In the wet fermentation process, microbes that have evolved to contain very active pectinase proteins are added to the beans. While this decreases the overall time from pulping to roasting and preserves some of the natural qualities of the bean, it also reduces the amount of fermentation-derived aromas that can be produced. The long fermentation time required by the dry method lets these other proteins add new characteristics to the bean. And, this is where *terrior* comes in again. Because the microbes in Africa are slightly different from the microbes in South America, this can have a profound effect on what the beans become and, differences aren't just seen between continents. Just like how the microbes in your armpit are different from the microbes on your nose, microbes can also vary from farm to farm and from region to region.

One question that comes up as a result of this description is: "How important are these additional flavor molecules?" To answer that question, I point you in the direction of Black Ivory Coffee, some of the most expensive coffee in the world.[14] This coffee contains beans that have passed through an elephant's digestive system and are harvested from its excrement. While the cost for these beans has a great deal to do with the difficulty in producing them, their mere existence shows the lengths that some will go through to produce new and interesting flavors in coffee.

At Compass, Brandon Warner and I discussed some of these unique flavors and how he approaches whether or not he preserves them through the roasting process. Warner tells me about this bean from Papua New Guinea that was really interesting. He did up several roasts of this bean, some lighter and some darker. He loved the light roast. Some of the nuanced aromas of the bean really came through. The brewed coffee was complex and unique. It was unlike anything he'd ever had before. The rest of the staff at Compass wasn't so sure though. They all liked a darker version of the roast instead. This might have more to do with expectations than anything else. The light brew didn't fit the profile of what was expected.

I asked him if they've ever done funky roast that they all loved. Overhearing this, Haft jumped in and he and Warner said, in unison, "Bali Kintamani". Warner said it has this really unique strawberry flavor that everyone was so surprised and enamored with. He also said that this version of bean is really difficult to source. They tried a lot of beans from Bali before they found this one. But, the staff all agreed that the highlighted uniqueness of this coffee was irresistible.

The interplay between identity and depth of flavor is a tricky one. Roasting completely transforms the bean in ways that are obvious to us. A fresh bean is a dull green color that smells of earthiness (grassy or like musty fruit). A roasted bean is brown and shiny and has an aroma that can fill a room. A package of freshly roasted coffee can make even the most ardent coffee-hater drool uncontrollably. When you roast coffee, you create molecules that we, as humans, recognize as being safe and beneficial for us to eat. As far as the chemistry of food goes though, the creation of new flavors and aromas often occurs at the expense of flavors and aromas that are intrinsic to that same food. Translated: roasting coffee beans can be detrimental to the flavors that are present before roasting.

This fact can manifest itself in different ways. The natural flavors can be overpowered by the newly produced flavors. Or, the natural flavors can be converted into new flavors through chemical reactions. In the case of roasting coffee, both of these happen.

The simple sugars that are found in coffee beans (remember how the dry fermentation method produces more of these) can go through both caramelization and the Maillard reaction. Caramelization, the reaction that makes caramel, is simply the reaction that occurs when you heat up molecules of sugar. In coffee, this reaction can result in flavors that are reminiscent of toast, nuts, and berries.[15] The Maillard reaction is another powerhouse for creating flavors and is probably the most important in terms of its effect on determining the final flavor of the roasted coffee bean. The Maillard reaction occurs when specific types of simple sugars are heated in the presence of proteins. The reason why Maillard is more important than caramelization

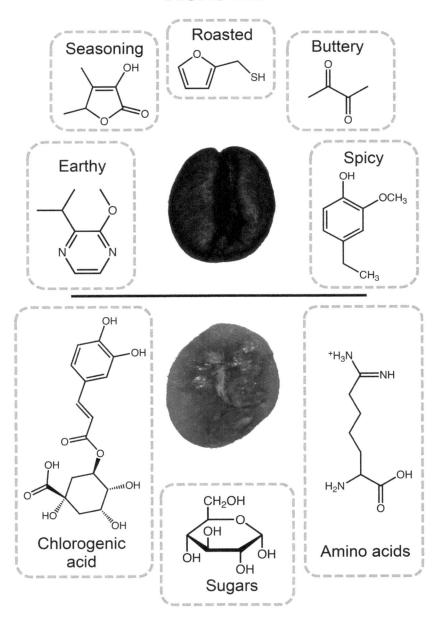

is that it can occur at lower temperatures. As the temperature of the beans is increased in the roasting process, Maillard reactions start occurring before caramelization reactions. In coffee, Maillard produces flavors that we recognize as roasted, nutty, potato-y, chocolate-y, and cabbage-y.[15]

For most of us, we don't think of coffee as being cabbage-y. I would even argue that most of us wouldn't think of coffee as being nutty. But, all of these aromas, added together at just the right amounts, result in what we recognize as coffee.

Much like the sugars and proteins, there are other important molecules that react away during roasting. Chlorogenic acid will turn into vanilla and other flavorings and, in the process of changing, will make the coffee less acidic. Caffeine will turn into other molecules as well. This is why lighter roasts are more acidic (sour) and caffeinated, while darker roasts are less so. This may come as a surprise to many coffee drinkers who think that stronger roasts are, well, stronger. It is true that they do have stronger flavors; longer roasts do tend to result in more acrid or bitter flavors. But they are most certainly less caffeinated.

Good control over this process couldn't be more crucial to consistently manufacturing the kinds of flavors that a roaster wants in their beans. Warner tells me that this 'consistency' is the biggest reason why their 30 kg roaster is so much better than working in a cast iron pot or even in their small batch roaster. As he takes me for a tour around this instrument, I can see why. There is a lever arm that is constantly mixing the beans into, and out of, the main heating area. There is a vent that pipes all of that wonderful aroma out into the neighborhood (perhaps this is the reason why so many people line up for their coffee). Finally, and most importantly, there is a computer screen that shows the control panel and readings from the roaster. Warner stresses that the readings panel is crucial to the consistency of their products. He tells me that, while the final temperature of the beans can give you an indication of the type of roast that you are going to get, the speed at which the beans get to that temperature is just as important.

And, sure enough, the panel displays a slope showing the increase of bean temperature during the roasting process. Warner tells me that this is a good batch and that the heating rate is matching what he had hoped. He opens up the file from an old run to show me one that didn't work; these beans reached their final temperature much too quickly.

Warner then brings me back to where they keep their green coffee beans. He shows me two samples that different suppliers are trying to sell them. The first set has noticeable divots taken out of them. Warner tells me that these are from insects that feed themselves on the fermenting beans. He notes that there is nothing inherently wrong with these beans and that insects are just a fact of life in this business. However, coffee beans that have been damaged by insects are going to roast at a different speed than ones that are undamaged. This damaged bean has a higher exposed surface area and lower internal volume. Because consistent quality is what Compass strives for, they need to have consistency in their coffee beans as well.

As Warner and the crew at Compass seem open to experimentation and finding really unique flavors, I push him a little farther on what they are looking for in a coffee. He says that a lot of what they make is traditional and good. They have a few restaurateurs who request specific flavor profiles. A BBQ restaurant gets a coffee that has extra smoky flavors in comparison to their other roasts. Normally though, every roast is an all-hands-on-deck decision. Warner will take a set of beans, roast it to different temperatures (light, various mediums, and dark), and the staff will taste test and comment.

But, all of this flavor creation isn't any good if you can't get those flavors to go into water during the brewing process.

FLAVOR EXTRACTION

So now that the coffee beans have been prepped, they need to be brewed. "What's the big deal with brewing coffee anyway," you might ask. Just add some hot water to the beans and filter, right?

And the answer to that is, "Well ... kind of."

The real answer to that gets us back to the idea of "perfection", which is really defined by whatever it is that you like best.

I have a quick confession to make. I only recently started drinking coffee. I plodded through the entirety of my PhD studies and postdoctoral work (years upon years of toiling away with experimental minutia, looking for new little details about nature that no one had ever observed before), without becoming addicted to coffee. This little bit of personal history may feel like sacrilege to a lot of my colleagues, whose lab training was fueled by gallons upon gallons of coffee. But, it was never really my thing. Only later did I find out that I really like espresso drinks. Regular coffee never did it for me. Along with that, my stomach was always out of commission for the day because of how the acid in coffee affected me. Only now do I know that the extra roasting needed to make espresso breaks down more of the acids that are naturally found in coffee beans. I still have a hard time drinking regular coffee. Like every other human, my personal preferences determine what "perfection" is. Just like me, your perfect perfection requires the right kind of roast and the right kind of brew.

Walk into any coffee shop and you'll see lots of options toward your perfect, right at your fingertips. Light roast. Medium roast. Dark roast. Seattle roast. French roast. Albuquerque roast. Leicester roast. French press. Pour-Over. Drip. Espresso. *Café au lait. Café Americano.* Quad grande, non-fat, extra hot caramel macchiato, upside down.

Perfection then, is part roast (flavor creation) and part brewing (flavor extraction).

Coffee, by one scientific definition, is an emulsion: a mixture of oil and water. We'll be discussing emulsions a lot in this book. There are too many experiences in life to list, that show us that oil and water don't like to mix. Vinaigrettes are probably the best example of this. Whether you make your own, or buy them at the market, vinaigrettes always end up as a layer of oil on top of a layer of water.

It turns out that there are things that you can do to make oil and water mix, at least for a bit. Give them enough time and they will separate. Even in coffee, after you let it sit out and cool long enough, you can see the oils separated from the water. If you look closely enough, you'll see iridescence, that rainbow-like pattern of colors, sitting on the top of your coffee. This is the same pattern that you'll see in a parking lot after it rains, when the spilt motor oil on the asphalt rises to the top of a puddle.

All of those flavor molecules that I described earlier would be much happier swimming around in oil than in water. The technical, chemical, jargon-y way of describing this would be to say that the separated water-flavor molecule system has a more stable energy than the system where the water and flavor molecules are interacting with one another. In chemistry, low energies denote more stability than high energies. The higher the energy is, the more likely

Coffee

it is that a system will fall apart. Whether that system is a single molecule, with its atoms or bonds; or an entire ocean, with water and salts and everything else that an ocean contains; an intact system with a lower energy than the separated system will remain intact, and the separated system with a lower energy than an intact system will eventually separate. To make this description a little clearer, sugar and salt dissolve, or mix evenly, with water because the water-sugar or water-salt interactions give the resulting syrup or saline a more stable (lower) energy than the system where solid chunks of sugar or salt are sitting at the bottom of a container of water. Water-oil mixtures on the other hand are less stable because they have a higher energy than oil that has separated from water.

To make oil and water mix, one has to add extra energy to make up for the difference between the stable (low energy) separated system and the unstable (high energy) mixed system. That energy can come from a lot of different sources. It can be mechanical. When you make a vinaigrette, whisking the oil and water is a form of mechanical energy. Or, that energy can come in the form of heat. The water molecules for your coffee are heated so that they extract the flavors (oils) from the grounds. As your coffee cools, the flavors naturally separate from the water, creating the iridescent slick on top of your once-tasty coffee.

Chemists describe molecules (and flavors) that have favorable interactions with water as being hydrophilic (water loving). Molecules that do not have favorable interactions with water are considered hydrophobic (water hating). Hydrophobic is kind of a misnomer. It's not that the molecules fear water or hate water or are repelled by water. It's that they just don't get anything out of being around water. Reminiscent of the "it's not you; it's me" line that you have surely used on an ex at some point in your life, you would just have rather been with someone else! And many flavor molecules feel the same way when it comes to water.

So what makes a molecule hydrophobic or hydrophilic? The easiest answer to that question is that hydrophilic molecules tend to have a lot of charges or –OH groups (an oxygen atom bonded – right next to – a hydrogen atom) on them. Hydrophobic molecules tend to contain a lot of carbon atoms.

"That's nice, Mr Chemistry Guy," you say. "I don't think in chemicals or have any memorized. How am I supposed to know when a molecule has an –OH or a charge or whatever?" You're not! I have lots of drawings of molecular structures of important flavor and food-related chemicals in this book. As I'm discussing one of these chemicals, I might refer you to its structure to highlight why it acts the way it does in cooking. For a chemist, these pictures are incredibly helpful in relating how it will interact (with you, with water, with oil, with other food molecules) or how it might react (and change into something new). And while these drawings may be new to you, they are the most minimal way of showing what a food molecule might do. As we go through this book, I will be your guide for pointing out which parts of different molecules are important. And, who knows, maybe at the end of this book, you'll want to learn more!

OK, so let's get back to the coffee flavor molecules. Specifically, let's talk about chlorogenic acid and whether it is hydrophobic or hydrophilic. Chlorogenic acid has 6 –OH groups. It also has 16 carbon atoms. (Carbon atoms can be shown by a "C" or can also be shown as a corner or angle. For example, there are two hexagons in the structure of chlorogenic acid. Each vertex, or corner, in that hexagon is a carbon.) So, chlorogenic acid has multiple –OH groups and multiple carbon atoms. Is it hydrophobic

CHLOROGENIC ACID

1 bond between carbon atoms

2 bonds between carbon atoms

Molecules have geometry.

Bold line = atoms above plane of paper.

Dashed line = atoms below plane of paper.

The OH groups have strong interactions with water

or hydrophilic? The answer to that question is kind of tricky. Chlorogenic acid is less hydrophilic than something like table salt (sodium chloride). But, and this is important for our current discussion, it is more hydrophilic than many of the other flavor molecules in coffee. And this comparison is what is important when we brew our coffee.

As the hot water soaks the coffee grounds while brewing, the water will start to wrest the flavor molecules from the beans. These molecules have a natural inclination to remain in the grounds rather than seep out into the water. However, with increasing temperatures of water, the heat energy from the water is able to compensate for the stability of the flavor molecules in the coffee grounds. At higher temperatures, the water will be able to pull out more flavor than at lower temperatures.

Furthermore, flavor extraction isn't an all-or-nothing, instantaneous affair. That is, the hot water doesn't pull out all of those flavor molecules at once. The flavors come in waves. The first wave to hit includes the most hydrophilic molecules. For coffee the most hydrophilic molecules tend to be acids, like chlorogenic acid. The most hydrophobic flavor molecules tend to include the bitter flavors and there tend to be more bitter molecules in coffee than acidic molecules. This is why over-extracted coffee, made when water that is too hot remains in contact with coffee grounds for too long, will always taste bitter: there are not enough acid molecules to balance out all of the bitter flavor. In the middle of these extreme waves of flavor are most of the flavors that we tend to associate with coffee: roasted, savory, caramel, buttery, and coffee-y. And, much like the ocean, these waves are rarely clean; they don't come into the shore on their own. Waves (or flavors) coalesce, overlap, and come into the shore (or your cup) at the same time. As the wave of acidic flavors is ending, the next wave of flavors is already starting to crash. And, the bitter flavors start to come through before the other flavors have been fully extracted.

So why not brew your coffee so that you only get the roasty flavors that you want? There are a couple of good reasons why that won't work. It's really difficult to isolate individual flavors from mixtures. There are trained PhD chemists whose sole job is to figure out what exact flavors are in coffee and how much of these flavors there are. This is no simple task. The other reason that

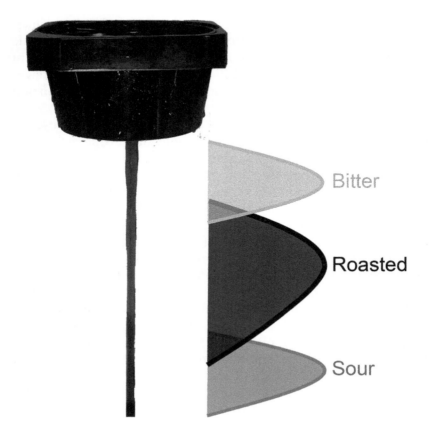

this is difficult is that coffee won't taste like coffee unless there's a little acid and a little bitter to it.

It's a complicated story.

The flavor of the coffee is definitely affected by the fermentation and the roast. But that black liquid in your mug is also affected by the amount of time between the roast time and when you brew it. You know those wonderful aromas coming from your coffee beans? Well, if you can smell it, it's not in your coffee anymore. Your cup of brew is also affected by the amount of grounds, the size of the grounds, the temperature of the water, and the amount of time the water is in contact with those grounds.

Your coffee maker at home and in the office was designed to handle any type of coffee that you put into it. If it were designed to work with only the freshest roasted coffee, it would operate differently. If it were designed to work with only the cheapest coffees, it would operate differently.

This is where those Pour-Oover coffees come in to play.

Now, before I get into any of that I need to add a disclaimer. Coffee drinkers come in all shapes and sizes and predilections. Some like theirs so long as it's hot and caffeinated. Others like their coffee weak. And some take it black like tar.

But, no matter how you take it, you can find a way to brew it.

David Andrews was a colleague of mine while I was doing my PhD at Northwestern. He was a few years younger on the same track. David and I both worked on developing different chemical theories to complement experimental research being performed in different labs at the university. David currently works for the Environmental Working Group, assessing chemical hazards that all people are exposed to during the course of our lives. To say that he and his wife, Lena Sadowitz, love coffee is a huge understatement. They have chronicled their search for "perfect" brewing methods on their website, "Not Your Parent's Coffee."[16] They test other brewers recipes using: Pour-Over, French press, Aeropress, Siphon brewing, and many others. They even develop their own recipes. One of their favorites, which uses Pour-Over techniques uses the following procedure: take 35 grams of coffee, ground to between espresso and drip sizes. Add this to a Pour-Over filter holder with a filter. (They even tell you which filters they like best.) Over 4 minutes, extract the coffee with 560 grams of water that is heated to 97 °C.

To many of you, this will seem overly fussy. But, to David and Lena, it's about making what they like and learning how to do that consistently. And, I suppose that is one of the things that this book is all about: talking about some of the most fundamental chemistry in everyday cooking so that the food we make meets with our ideas of perfection. I do hope that the chemistry of cooking enhances your enjoyment of eating and being in the kitchen, much as it has enhanced mine.

REFERENCES

1. S. Ponte, The 'Latte Revolution'? Regulation, Markets and Consumption in the Global Coffee Chain, *World Dev.*, 2002, 30(7), 1099–1122.
2. M. R. Hartings, I. Love Gin and Tonics, The Best Science Writing Online 2012, *Sci. Am.*, ed. J. Oulette and B. Zivkovic, Farrar, Straus and Giroux, New York, 2012, pp. 143–147.

3. M. R. Hartings and D. Fahy, Communicating Chemistry for Public Engagement, *Nat. Chem.*, 2011, **3**, 674–677.
4. *Ratatouille*, Hollywood, Pixar, 2007, Film, directors Brad Bird and Jan Pnkava.
5. L. M. Principe, *The Secrets of Alchemy*, The University of Chicago Press, Chicago, 2013.
6. N. Myhrvold, C. Young and M. Bilet, *Modernist Cuisine*, The Cooking Lab, 2011.
7. M. Colonna-Dashwood and C. H. Hendon, *Water For Coffee*, Bath, Colonna-Dashwood and Hendon, 2015.
8. *Espresso Myths Debunked*, Chef Steps. [Internet site. Cited June 2016], available from https://www.chefsteps.com/activities/espresso-myths-debunked.
9. M. Ravindranath, *At Compass Coffee, Data is the Secret Ingredient*, The Washington Post, 2014.
10. L. M. Jarvis and J. Morrison, Nitro Cold Brew, *Chem. Eng. News*, 2015, **93**(33), 37.
11. F. Denoeud, *et al.* The Coffee Genome Provides Insight into the Convergent Evolution of Caffeine Biosynthesis, *Science*, 2014, **345**, 1181–1184.
12. C. Zimmer, *How Caffeine Evolved to Help Plants Survive and Help People Wake Up*, New York Times, 2014.
13. L. W. Lee, M. W. Cheong, P. Curran, B. Yu and S. Q. Liu, Coffee Fermentation and Flavor – An Intricate and Delicious Relationship, *Food Chem.*, 2015, **185**, 182–191.
14. M. Sullivan, *No.1 Most Expensive Coffee Comes From Elephant's No. 2. All Things Considered [radio broadcast]*, Washington, NPR, 2014.
15. P. Semmelroch and W. Grosch, Studies on Character Impact Odorants of Coffee Brews, *J. Agric. Food Chem.*, 1996, **44**, 537–543; F. Mayer, M. Czerny and W. Grosch, Sensory Study of the Character Impact Aroma Compounds of a Coffee Beverage, *Eur. Food Res. Technol.*, 2000, **211**, 272–276.
16. *The How to Make Coffee Database*, [Internet site. Cited June 2016], available from http://notyourparentscoffee.com/how-to-make-coffee-database/.

CHAPTER 2

Bacon

It's not everyday that a scientist gets a phone call from the *Today Show*.

I got the call as I was sitting in my office on an unseasonably hot and humid afternoon in May. I had spent a long morning setting up reactions in my lab and doing a bit of a clean-up. My phone started ringing just as I stepped back into my office for a break. The person on the other end of the line said that they were from the *Today Show* and would I like to go on camera with them to talk about the science of bacon.

I must admit that this request came as a bit of a shock to me. While the people closest to me know that I know a lot about bacon, I'm not necessarily recognized as a bacon expert.

The producer said that he had gotten my information from the American Chemical Society (ACS). He had contacted the ACS because he was interested in a video that they had made about the aroma of bacon.[1] The group that made this video, ACS Reactions, puts together all sorts of fun and informative videos about chemistry. I had worked with them in the past on separate videos about making ice cream[2] and grilling onions.[3] They are pretty amazing at what they do. Their videos

---

Chemistry in Your Kitchen
By Matthew Hartings
© Matthew Hartings, 2017
Published by the Royal Society of Chemistry, www.rsc.org

are always quick to grab your attention, and they manage to showcase some esoteric chemistry in a way that is very easy to understand. Because of that, people actually seek out their videos. I have been formally involved in scientific outreach and communication with non-scientists for a long time. It is not an easy task to get non-chemists interested in chemistry. While a non-scientist will purchase books on string theory and evolution, books on chemistry haven't had such a broad success. All of this is to say that the ACS Reactions crew is very good at what they do.

In this instance, they had made a video about bacon aroma. One would be hard pressed to find people who don't like the aroma of bacon. And the Reactions crew didn't just make a video saying that bacon smells good. Well, they did say this. But they said it in the context of talking about a research article that determined the exact compounds that make bacon smell like bacon.[4] The types of experiments, performed by María Timón and her colleagues in Badajoz, Spain, that figure out what aroma molecules are present in bacon, are no small feat. As with coffee, if you are smelling molecules wafting from bacon, those chemicals are no longer in bacon. Aroma is a fleeting property. And capturing those aromas is hard enough, even before a scientist tries to figure out what molecules they have captured.

But, the masses do love bacon. And, of course the ACS Reactions video went viral. Coupled with this, a chemistry teacher in the UK, Andy Brunning had created an infographic on the same research, which had become quite popular.[5] (Brunning has written an excellent book of chemistry infographics, which I played a small role in reviewing, called *Why does asparagus make your wee smell.*[6]) The video and the infographics had such a large audience that larger media outlets came calling, wanting to know all about the research. The good people at ACS Reactions could handle most of the calls. But, when *The Today Show* contacted them, they wanted to speak to a scientist who they could put on camera ... within the hour. Seeing as how the pertinent researchers were in Spain, and Brunning was in the UK, and none of the Reactions producers are scientists, they needed to find someone who happened to

Bacon

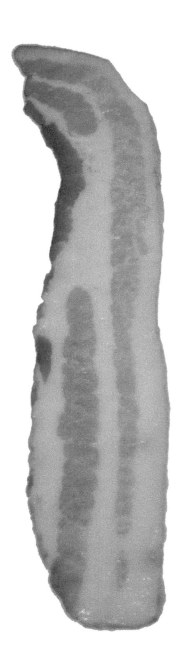 

be a scientist, knew a bit about bacon, and could speak coherently about María Timón's research.

For whatever reason, my friends at the ACS decided to pass my name on to the NBC production crew. Now, I do have some known credibility in this area. I created a chemistry of cooking class for non-science majors at American University.[7] I had worked with the ACS on other food science videos. I have written about the Maillard reaction, the chemical reaction that turns bacon from pink and floppy, to crispy and brown.[8] I also just happen to have family who raise hogs.

SOME PIG

I am sure that you are all wondering the same thing. And the answer is, "Yes. Yes, I have shown pigs at the county fair."

When I was a kid, several of my aunts and uncles raised pigs. Connie and Ralph, Diane and Denny, Mike and Pam, and Jane and Jerry, all had pigs on their farms. Even my uncle Mark, who is a Catholic priest, still finds the time and a place to raise a few hogs every now and then.

One of the great things about having so many aunts and uncles is having even more cousins. Seriously. I have 63 first cousins. There were always enough of us to field some pretty competitive basketball and (American) football games at family events. Within the group of us, we had "city" cousins and "country" cousins. The "country" cousins all grew up on farms. Any of us who didn't grow up on a farm were automatically labeled a "city" cousin. Even though I grew up on a wannabe, part-time farm, I was totally a "city" cousin. And, I am completely fine with that distinction. The amount of work and the actual time that my "country" cousins had to spend tending to their animals or the fields completely outweighs anything I ever had to do. Farming, for many, is as much a labor of love as it is a way to earn money. And, I have too much respect for what farmers actually do than to actually think that I was a farm boy, myself.

During the summer, I would sometimes get to go to stay with my cousins for a couple days. Every now and then, I would have to help out on the farm. But, the visits I remember most were when I would go during the county fair. There was something utterly freeing to be able to run around with my cousins all day at the fair. Mom and dad would send me with a little bit of money, and we would eat fair food and ride the rides. One memory that particularly stands out to me, is when I rode the Cyclone with my cousins Kyle and Andy. Waiting in line, we had been fighting over who would get to sit in the outside seat. When they put us in our car for the ride, they sat me on the outside because I was the tallest. Andy and Kyle were furious; older than I was, they felt like they had somehow been deprived of their natural right to sit in the outside seat.

The other thing that I would get to do at the fair is show some of my cousins' animals. They would have plenty of animals to show: cows, pigs, and chickens. I only ever got to show pigs. Pigs were much easier than chickens, which had to be held, or cows, which could become uncontrollable if you didn't handle them right. In other words, us city cousins could be trusted with showing a hog. My cousin Dan, another city kid, managed to win a ribbon one year, besting several country cousins in the process. If I'm not mistaken, Danny still counts this as one of his major achievements in life. (For the record, he is married to a wonderful woman, is a father, and is a successful lawyer.)

So, it should be rather apparent that my family deeply appreciates, and are proud of their livestock.

Despite this, I find that I'm always learning more about how farm animals are, and can be, raised. My Twitter stream is partially populated with food scientists, food ethicists, professional chefs, food and agricultural journalists, and farmers. These people, passionate about their work, often find themselves in conflict with one another.

Although they have a shared goal: feeding all 9 billion humans and counting; the complex problems surrounding this goal have no easy answers. Food scientists tend to look to technologies that can increase crop and livestock yields and deliver foods free of pathogens that can stay fresh for a long period of time. Food ethicists help us to grapple with balancing concern for how we treat animals and land and the environment, with the need to sustain our growing population. Professional chefs are a reflection of how our culture prepares food and often transforms that same culture in their efforts to expand and enrich our culinary experiences. And, farmers have the toughest task of all of these. Not only do they have to put in the manual labor required by the animals and plants that they tend to, they have to balance the demands and suggestions and technologies developed by the scientist, ethicist, and chef.

It is easy to see where all of their points of view come into conflict. The ethicist sees overreach. The scientist feels restraint. The chef finds imperfections. And the farmer feels pulled in too many directions. But, for all of their disagreements, all of these professionals are trying to sustainably and effectively bring us food that will nourish and enliven us.

The job of the journalist, then, often lies in keeping all of these professionals honest. Maryn McKenna's reporting, amongst others, has made us aware of the overuse of antibiotics in farming.[9] While the use of large quantities of antibiotics can optimize the survival rate of livestock, from birthing to butchering, lowering the cost of meat at the grocery, overuse can also facilitate the evolution of resistant bacteria leading to untreatable infections. Brooke Borel has written about the fine line facing researchers working at land grant universities.[10] The mission statement for these institutions is to support farmers who want to use new agricultural technologies. The individual researchers though, fall under extra public scrutiny whenever they work with the corporations that have developed those same technologies. If farmers want to use new technology, that technology most likely comes from some player in the agricultural industry, such as the often-maligned Monsanto. Researchers at land grant universities must perform their duties and assist the farmers. However, if the scientists are viewed as working too closely with a corporation, they can easily and quickly lose the trust of the public, who ultimately supply the funding and support for that very work.

In this same vein is Barry Estabrook's *Pig Tales*.[11] It is at once a story about how difficult it is to raise pigs and prepare them for butchering and deliver them to our markets as affordable cuts of meat. It is also a story about how some farming practices lead to shortcuts that compromise quality and the wellbeing of the animals; farming remains a delicate balance. Estabrook has a few suggestions, but many of them fall short or, by my reading, lack real conviction for sustainable changes to the system in a way that is affordable for the average consumer in both developed and developing parts of the world.

Most of all, though, his book is a love story about pigs. They are incredible creatures, who have played a role in nourishing us, as well as advancing our culture. They have the power to decimate ecosystems. They have the intelligence to solve complex problems. They really have changed us. And we, in turn, have changed them.

PIGS ARE MAGIC

Yes, pigs are magic, as the saying goes. What other animal do you know of that can turn vegetables into bacon?! And, please don't say turkey. Turkey bacon really doesn't hold a candle to the real deal.

For all of their magical powers, humans play as much of a role to produce the prosciutto, fashion the *jamón*, and bring home the bacon.

Human intervention comes in all shapes and sizes. Agricultural development resulted in pigs that could be kept on farms. These "tamed" pigs were bred and cross-bred to produce new varieties with the right temperament, the right type of meat, and the right natural flavoring. All of this takes time: tens, hundreds, thousands of years. Patience, perseverance, and a lot of luck helped to give us the animals that we recognize today.

Intervention is also immediate. Pig feed is controlled so that the meat has the right amount of marbling (ratio of muscle to fat and how the two intermingle). In extreme (and extremely tasty) cases, pig feed is chosen to completely dictate the flavor of the meat. The most famous example of this is *jamón Ibérico de bellota* from the black hooved Ibérico pigs that only feed on *bellota* (acorns) that fall on the pastures where they live.

But, the most profound effect that we have on bacon is in the way we prepare it for eating; we salt it, we smoke it, and we fry it. Each step adds a different layer of flavor, without which bacon just wouldn't taste the same. Salting pork has an obvious effect on the flavor along with several subtle changes to the texture of the meat. Smoking imbues the meat with a multitude of aromas and a hint of savory spice. What I'd like to focus on, though, is what happens when we fry bacon.

Before I even describe it, you can hear it, and smell it, and your mouth waters. We have a very Pavlovian response to frying bacon. The sizzle of the fat in the pan. The aroma that infuses the air, only to dissipate and disappear hours after it has been cooked. It is unavoidable, a part of the human condition. We are evolutionarily predisposed to loving bacon and seeking it out.

Richard Wrangham, an anthropologist at Harvard University, is the author of a book titled *Catching Fire*.[12] In this book, he posits that humans became human because our ancestors started cooking their food. The theory runs as follows. Digesting food is hard work. Our evolutionary cousins: chimpanzees and bonobos and gorillas and orangutans, spend a good portion of their day chewing and eating and eating and chewing. And, when they're not eating and chewing, they are sleeping while they digest all of that food that they've eaten and chewed. All animals need energy to live. To get that energy, we eat food. But, to get the energy contained in the food, our bodies have to break the food down: turning big molecules like proteins into little molecules like amino acids. Digestion requires us to spend our own energy. So these apes have to spend a lot of energy to extract a little more energy so that they can live and mate and take care of their children. (The parents or caregivers reading this understand how much energy is required to raise a child.) For these apes, eating and digestion is exhausting but necessary.

When those early humans started throwing meat into fire they got an unexpected benefit. It turns out that cooking is effectively a kind of digestion. The partial digestion, provided by cooking, made it easy for these people to get energy out of their food. Eating was no longer so exhausting. Because they didn't have to use up all their energy in digestion, they had extra calories to burn. Those extra calories, Wrangham postulates, went to developing higher brain functions. Cooking, it seems, it what separated us from the animals.

MAILLARD IS MAGIC TOO

The effect of cooking on easing digestion is, no doubt, important. There is a chemical reaction though, something more carnal though no less intricate, that taps into our base desires and makes us love bacon so much. As soon as you throw meat into a fire, or put bacon on a hot pan, those aromas that immediately make themselves known are produced as a result of a process called the Maillard reaction. Frying bacon, searing a steak, baking bread, toasting bread, baking cookies, roasting coffee, and roasting meat all involve the Maillard reaction. The crust you get on your steaks or bread while they are heated come from the Maillard reaction. Caramelizing onions doesn't actually caramelize them, it Maillard-izes them. The browning that occurs as you make these foods, all come from the Maillard reaction. All of the incredible and savory and irresistible aromas and flavors that develop while cooking these foods are also a result of the Maillard reaction.

Our bodies, it seems, are specifically tuned to recognize these flavors and aromas. I already brought up the unmistakable smell of frying bacon, but the smells produced by the Maillard reaction in other foods are just as mesmerizing. The aroma of baking bread is instantly recognizable. Sometimes I think that my wife, Erika, and I bake our own bread just to make the house smell like a bakery. We (me, you, your mother, *etc.*) can tell when chocolate chip cookies are about ready when we can start to smell them from the oven. We can even quickly recognize the Maillard reaction in something as simple as making toast. It is quite obvious that this reaction is important for much of our cooking, and that we, as humans, are rather enamored with it.

The prevalence of the Maillard reaction is due to the fact that it occurs in foods that contain both proteins and sugars. As that includes most of the things we eat (and cook), the Maillard reaction is inescapable.

The Maillard reaction is the set of chemical transformations that takes place when proteins react with reducing sugars and is named after Louis-Camille Maillard. In the early 1900s, he observed the formation of a yellow color when sugars were mixed with amino acids (the chemical parts that make up proteins) and heated. The eventual fame and importance of this reaction came after Maillard's death.

The "reducing" qualifier attached to sugar is an important one. It tells us that sucrose (table sugar, which is made up of a glucose molecule chemically attached to a fructose molecule) cannot be involved in the Maillard reaction, while glucose and fructose, on their own, can be.

If we think again of the chemical structure, as chemists are wont to do, we can understand how a sugar becomes "reducing". Sugars like fructose and glucose are often imagined and drawn as being in a ring structure, where either a hexagon (in the case of glucose) or a pentagon (in the case of fructose) are the defining characteristics. When you are holding a bit of glucose powder the molecules pretty much all have this ring structure. But, as soon as you throw glucose into water some of those rings open up. Not many do. Eventually, all of the individual glucose molecules will be "open" at one time or another. The rings open and close and open and close. Only about 1% of the glucose molecules in water are "open" at any time. Importantly, at the end of one of these open glucose molecules is a chemical group called an aldehyde (a carbon attached to an oxygen, through a double bond, and to a hydrogen, through a single bond). This chemical group is called "reducing" because it has a propensity to be oxidized (the chemical inverse of reduction) into a carboxylic group. In this portrayal, "reducing" is a chemist's shorthand for saying that electrons will eventually move away from the carbon in the aldehyde. Implicit in the sugar's description as "reducing", is the expectation that the sugar will reduce a different molecule. The details of this process, for the moment, are not important for our description of the Maillard reaction. What is absolutely necessary, though, is that the sugar ring opens, presenting the aldehyde group at the end of the molecule.

At the right temperatures, when a reducing sugar is near an amino acid, especially an amino acid with an amine group (a nitrogen atom that is attached to a carbon and two hydrogen atoms), these two molecules will transform into new chemicals as they go through the Maillard reaction. There are a number of molecular changes that occur during this reaction.[13] The nitrogen from the amino acid makes a bond with the aldehyde carbon of the reducing sugar. This initial step causes a cascade of changes in which the electrons that normally hold the molecule together, start bouncing around from one atom to another. All of this jumping can sometimes cause the modified molecule to lose a few atoms (often in the form of a water molecule), and can sometimes cause the modified molecule to gain new atoms (often from a water molecule). There are 5 unique changes that can happen, and all involve the movement of electrons and reorganization of the atoms around these movements. Compounding this complexity is the fact that all of the changes are in an equilibrium. What this really means is that the identity of the molecule going through the Maillard reaction is constantly changing from the starting molecules to one of the 5 other sets of molecules that have been altered. In a reaction between a single sugar and a single amino acid, the molecule that actually exists at a single point in time is randomly changing from one grouping to the next.

Maillard Reaction

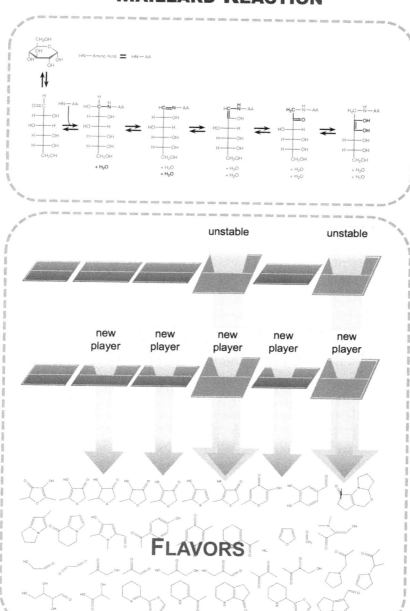

To understand how an equilibrium works, let's consider a single chemical change, one molecule turning into a second, which can then turn back to the original. You can think of this equilibrium as a game of table tennis with each side of the table corresponding to one of the molecules at either end of the equilibrium. The arrows that we use to show that change can occur in either direction make up our net. In a game of ping pong, the ball is constantly bouncing from one side to another, just as the identity of the molecule in the reaction is shifting from one to another. And, just as some players are quick to return a volley and some are slower, chemicals in an equilibrium tend to stay on one side of the reaction longer than they stay on the other.

Now we need to expand this analogy from a single reaction to the set of reactions we have in the Maillard process. If you imagine, then, that a second table is attached to the first and that the ball can be hit one of two directions. In one direction, the ball would just return over the net from where it came. In the second direction, the ball could be hit away from the first net and over the second net. In our extended analogy for the Maillard reaction, there are five nets and six unique places where the ball can be hit. At any time, the ball could just be bouncing back and forth over a single net. But, it could also be continuously sent over a first net and then a second net and then the third net and then ... The take home point to this is that, once the reaction starts the identity of the molecules (the ball's location on the extended table) is completely dependent upon how the game has been played.

At some point, this perpetual game of ping pong must end. Each Maillard reaction will eventually progress to one of two outcomes: a small flavor/aroma molecule, or a larger molecule that becomes part of the crust/crispiness or browning color. To get from your game of chemical table tennis to the flavor or color molecules, two different things could happen: (1) one of the equilibrium molecules could irreversibly change into a new compound that will break down further to form one of the flavors; (2) some other amino acid or sugar could get in on the act and force its way into the reaction, which will ultimately lead to the formation of a flavor molecule or a larger molecule that adds color or crispiness.

Going back to our table tennis analogy, the game (reaction) ends when the ball lands on the floor. There are different ways that this can happen. The ball can just bounce off of the table on its own (scenario 1 above). Some tables might be more tilted than others, just as some molecules are more prone to decomposition than others. For the second scenario, a new player could join in and hit the ball to the ground on her own.

If you were able to run just a single Maillard reaction, you would get only one of these final products (either aroma or color). The identity of the molecule that gets produced is a function of where and how it leaves the table. That is, a tilted table (unstable molecule) will create a different outcome to that created with the entry of a new player into the game.

The aromas, chopped up portions of the original reactive molecules, include things like pyrazines (toasty flavor), pyrroles (nutty flavor), alkylpyridines (bitter or burnt flavor), acylpyridines (cereal-like flavor), furanones (sweet, caramel-like flavor), furans (meaty flavors), oxazoles (green flavors), thiophenes (meaty, roasty flavors), and many others.

Within the second scenario, the identity of the new player also has implications for what the resulting molecule ultimately becomes. To drive home this final point, Maillard reactions for grain products (bread, beer, *etc*) tend to have similar smells because these foods contain more of the amino acid, proline. Proline can force its way into the equilibrium reactions to generate new flavor and aroma molecules. Meat tends to contain more of the amino acid, lysine, which will force its way into the equilibrium reactions to generate different flavor and aroma molecules. Baked grains have a different aroma than cooked meats precisely because of the fact that grains have more proline amino acids than meats, which have more lysine amino acids.

We know that a single Maillard reaction (between a single reducing sugar and a single amino acid) will give a single set of reaction products. If you run the reaction one hundred times, there is a chance that you would get one hundred different sets of compounds out at the end. When we eat foods that have gone through the Maillard reaction, we experience a complex mixture of flavor and aroma molecules. When you run the Maillard many trillions of times (like you are doing when you cook food), then you start to statistically build up that group of aromas and flavors, some molecules being more probable outcomes of individual reactions than others. Running the reaction many millions of times, you find that some molecules are more likely to be made than others. This has to do with a number of factors: the temperature, the slant of the ping pong table (how stable different reactive molecules are), the number of other players that try to enter the game (the concentrations of the other amino acids that are present during the reaction). When you run these reactions enough, you start to get a good understanding of which flavors are going to dominate what you are cooking.

The same is true for scientists who study the Maillard reaction. We have a pretty good idea of the first 8 reactive compounds that are present in the complex equilibrium. We also have a good idea of the identities and proportions of different aroma and flavor molecules that get made in the end. The pathway from the complex equilibrium to the final product is much trickier to study. In research chemistry we aim for simplicity by trying to control and observe a single reaction pathway. It is difficult to dictate which part of the complex, 5-step equilibrium initiates the reactions that lead up to a different flavor molecule. And, as you can surmise, if you generate several hundred unique molecules, there are, at least, several hundred unique reactions that lead to those molecules. It is a fascinating chemical process and one that continues to stump chemists. In many ways, we are much more adept at running and controlling this reaction in the kitchen than we are to running and controlling this reaction in the laboratory.

FOOD SAFETY AND EVOLUTION

Getting back to Richard Wrangham's argument about human evolution and cooked (more easily digested) food, I'd like to talk about how I think the Maillard reaction plays into this our development as a species.

The fact that the Maillard reaction occurs in our food is by the mere coincidence that most of our food contains proteins and sugars. There are no chemicals produced by this process that immediately stand out as things that should be pleasing to humans. Take pyridine, for example. Pyridine is a small amine compound. One quality shared by many amines is that they smell horrible. Putrescine and cadaverine fall into this category, and their names suggest all you need to know about how they smell. Pyridine is no different. I work with it in my research lab. It smells like death. It is offensively odorous. Even when I work with it in a chemical fume hood, which protects me from its stench, I always get nervous that I might get a small drop of the stuff on my clothes. Once that happens, you are better off just burning your clothes than trying to wash them.

Pyridine, along with several related molecules, is made during the Maillard process. But we don't experience the aroma of pyridine on its own in this context. When we smell frying bacon, we experience pyridine within a panoply of other compounds. It is this panoply that we find so appealing. Remove pyridine, or one of the other molecules from the mixture, and the mixture doesn't quite smell the same. Pyridine, this foul smelling molecule, is necessary to make fried bacon smell so good! For most foods that we cook and eat, the experience is an integrated event. What this means is that the taste, aroma, texture, appearance, and sound are all critical to how much we enjoy our meals.[14] The same is true for flavor and aroma. Aroma is an integrated experience. Our bodies recognize the mixture of chemicals that come from frying bacon, by sensing each molecule and tying our reaction to this mixture into a unified experience. All of the Maillard molecules, experienced at the same time, are important for our appreciation of a meal.

Again, Wrangham posits that cooking food was essential for human development because of the facilitated digestion that cooking offers. But, that is not the initial reason (or initial benefit) that our early ancestors would get from throwing meat onto a fire. The immediate and most pronounced benefit of cooking food comes from killing off any pathogen (bacteria, viruses, fungi) that might have infested the surface of meat. In our current age, we have a continuous system of chilled processing, transport, and storage that keeps our food fresh and un-spoiled. Our ancestors were always at risk of becoming ill from eating tainted food. Throwing that food in the fire, however, would immediately diminish this risk. Hominids who cooked their food had a better chance for living longer lives, birthing and raising more offspring. The ones who would recognize cooked food as being pleasing (to smell and taste) were more likely to put their food into the fire. Those who enjoyed and sought out cooked food would pass on their traits for liking cooked food to their offspring.

If you recognized cooked meat as tasty, you were more likely to live longer and produce more offspring.

If you recognized the molecules that form when you cook meat as tasty, you were more likely to live longer and produce more offspring.

If the Maillard reaction products appealed to you, you were more likely to live longer and produce more offspring.

If the Maillard reaction products appealed to you, chances are good that they would appeal to your offspring as well.

Our love of the Maillard reaction is written in our genes.

Our Pavlovian response to the aroma of frying bacon was born millennia ago.

The ability of cooked meat to assist in digestion might be the reason why we eventually evolved. But, we, as a species, would have never found this out without first developing a love of the chemicals that are produced by the Maillard reaction.

MY FIVE SECONDS OF FAME

If the preceding pages are an indication of anything, it is that I have thought about bacon ... a lot.

So, when I got the call to go on TV, I was ready.

Or at least I thought I was. I got the call while I was at work, and NBC wanted to film the segment within the hour. I started freaking out. I had been toiling in lab all day and was wearing ratty jeans and an old T-shirt. (This outfit is a near necessity in a chemical lab. Even when wearing a lab coat, an acid spill or a drop of the wrong chemical on your clothes can really ruin whatever it is that you are wearing.) I was certainly not about to give up a shot to go on national broadcast television to talk about bacon science. No way! So, I started rushing around campus asking my colleagues if they had a dress shirt that I could borrow. It was the middle of May and it was unseasonably hot and humid in DC. And, if the prospect of going on national television wasn't enough to make me sweat, the unexpected cardio workout I was getting certainly was. Thankfully, I found a shirt that was presentable. I got myself dried off. And I headed over to the local NBC studios, which are about a block away from my college campus.

During the lead-up to my interview I kept worrying about how I was going to talk about the research. Even with a topic as drool-worthy as bacon, I really wanted to give credence to Timón's research and to the reason why we do like bacon so much. I needed to be engaging and certain and give just the right amount of technical information. Surprisingly, during all of my running, I had come up with a plan that centered on the Maillard reaction and our evolutionary past.

The producer from the local NBC affiliate put me at ease right away. We briefly chatted about the process and what they wanted. He was going to ask me the same three questions in about twenty different ways. I would spit out answers and they would pick the quote that they liked the best.

And then we started.

It was kind of off-putting getting the same questions over and over. But I felt like I was holding my own. I talked about how much I love bacon. I talked about how difficult it is to make the measurements that Timón and her colleagues made. I talked about how the molecules that they found in bacon, pyrazines, pyridines, and furans, show up in other foods that we eat. I talked about how, on their own, these molecules smell quite different to what you would imagine. They smell like other things altogether. (Remember, pyridine smells like death!) But, what makes bacon smell amazing is the exact combination of all of these molecules. Take away pyridine, which smells like death, and bacon doesn't smell like bacon anymore. Almost all of those molecules are necessary.

Finally, I talked about why it is that so many people love bacon. If you think about it in terms of human evolution, you get to understand how we are naturally predisposed to loving bacon. For our hominid ancestors, food became safe to eat if it was cooked, smoked, or salt cured. Doing any of these things would kill off bacteria or other pathogens that were growing on the meat that these early people had hunted for. A lower exposure to pathogens meant a longer lifespan. Hominids who recognized cooked, smoked, and cured food as being tasty, were more likely to successfully produce offspring. Because bacon is salted, smoked, and fried, it would seem that a love for this food is written in our genes.

Our DNA has rendered bacon irresistible.

At that point in the interview, I've got to admit, I was pretty impressed with myself. I was confident. I was being witty. I was sure I was giving them good material. After it was all over, I had blabbed about bacon for a half of an hour.

I left feeling pretty good and was curious to find out which of my brilliant quotes they would use.

The morning of the show's airing came, and I was a little excited. Even though I hadn't gone to the New York studio, I was going to be on broadcast TV. Along with that, the bacon bit was going to air right before a segment with John Malkovich. I convinced myself that since I've shared air time with Malkovich, I now officially have a Bacon-number[15] ... appropriate given the topic of my interview.

The bacon segment started with a clip from the ACS video.[16] After that, I came on screen. Crazy. My name along with my title: Assistant Professor of Chemistry, American University was there, too. I looked official. And then I started to talk.

"I love bacon. It really is an amazing food. The flavors that we get from frying really make us go crazy for it."

A half hour of eloquence in that interview, and the quote that they use is, "I love bacon."

"NEWS FLASH: Science says bacon is delicious. Story at 9!"

While I was giddy to get as much screen time as I did, I was upset that they didn't use any of my well-thought out material. I wanted to show the world that I could be a legit bacon scholar and instead I blather, "I love bacon."

My friends and family, however, had a different response. They loved it. And, in fact, I still get people asking me to talk more about bacon science. Some of my in-laws have even become fond of calling me "Bacon Boy."

What I've realized in looking back on this, is that the producers (and my friends and family) really wanted to hear about my passion for bacon. After I said, "I love bacon", they had all the material they needed.

If *The Today Show* was my 5 seconds of fame, overall, I'm pretty happy with it. I mean ... who wouldn't want to be known as Bacon Boy.

REFERENCES

1. *ACS Reactions, Why does Bacon Smell So Good*, [video file] 2014 May 27 [cited 2016 July 5], available from: https://www.youtube.com/watch?v=2P_0HGRWgXw.
2. *ACS Reactions. Ice Cream Chemistry*, [video file] 2014 June 17 [cited 2016 July 5]. available from: https://www.youtube.com/watch?v=-rlapUkWCSM.
3. *ACS Reactions, Fast Caramelized Onions*, [video file] 2013 July 9 [cited 2016 July 5], available from: https://www.youtube.com/watch?v=npZosJvE1nU.

4. M. L. Timón, A. I. Carrapiso, Á. Jurado and J. A. van de Lagemaat, Study of the Aroma of Fried Bacon and Fried Pork Loin, *J. Sci. Food Agric.*, 2004, **84**, 825–831.
5. A. Brunning, *Why does Bacon Smell So Good? – The Aroma of Bacon*, [document on the internet]. Compound Interest, 2014 April 16. [cited 2016 July 5], available from: http://www.compoundchem.com/2014/04/16/why-does-bacon-smell-so-good-the-aroma-of-bacon/.
6. A. Brunning, *Why does Asparagus Make Your Wee Smell?: And 57 Other Curious Food and Drink Questions*, London, Orion Publishing, 2015.
7. L. K. Wolfe, Kitchen Chemistry Classes Take Off, *Chem. Eng. News*, 2012, **90**(36), 74–75.
8. M. R. Hartings, Maillard Reaction, *Chem. Eng. News*, 2011, **89**(47), 36.
9. M. McKenna, *Superbug: The Fatal Menace of MRSA*, Free Press, New York, 2011.
10. B. Borel, *Seed Money*, Buzzfeed [article on the internet]. 2015 October 19 [cited July 5 2016], available from: https://www.buzzfeed.com/brookeborel/when-scientists-email-monsanto?utm_term=.dmzvjxJrKd#.joZj6VA3O0.
11. B. Estabrook, *Pig Tales: An Omnivore's Quest for Sustainable Meat*, W. W. Norton & Company, New York, 2015.
12. R. Wrangham, *Catching Fire: How Cooking Made Us Human*, New York, Basic Books, 2009.
13. F. Ledl and E. Schleicher, New Aspects of the Maillard Reaction in Foods and in the Human Body, *Angew. Chem., Int. Ed.*, 1990 June, **29**(6), 565–706.
14. C. Spence and B. Piqueras-Fiszman, *The Perfect Meal: The Multisensory Science of Food and Dining*, John Wiley & Sons, Oxford, 2014.
15. *Six Degrees of Kevin Bacon* [document on the internet], wikipedia 2005 July 12 [cited July 5 2016], available from: https://en.wikipedia.org/wiki/Six_Degrees_of_Kevin_Bacon.
16. *The Science of Why Bacon Smells So Good*, The Today Show. 2014 May 29, available from: http://www.today.com/video/today/55285278#55285278.

CHAPTER 3

Eggs

My mom makes scrambled eggs better than you do.

I do apologize. But any arguments that you think you might have against this statement are invalid.

For me, scrambled eggs have a strong connection with Sunday mornings. On Sundays, whenever Mom would make brunch after going to church, she would always make scrambled eggs. There was nothing fancy about them, no fancy ingredients. She didn't source cream or cheese from the best dairies. She didn't use free-range, locally grown eggs.

Well, that's not an entirely true statement. There was a time when we raised chickens at home. My mom is a farmer's daughter. And there were several times during my childhood where we would have various kinds of livestock on our little parcel of land. Cows, horses, sheep, and chickens. I remember our time raising chickens with not a hint of fondness. Chickens are foul. Raising chickens was a dreadful period of my life. Being the oldest child, I was always tasked with doing the real chores. I would have to go and collect eggs every morning. In and of itself, this doesn't sound bad. However, the coop where they had their nests and laid their eggs was about 2 meters long by 2 meters wide and also contained a very territorial rooster. That bird would come after me with his talons and his beak whenever I went to get the

---

Chemistry in Your Kitchen
By Matthew Hartings
© Matthew Hartings, 2017
Published by the Royal Society of Chemistry, www.rsc.org

eggs. Of course, I can't blame the rooster for doing so. It's not as though I was doing something that he liked. Whatever empathy I might have harbored for him wasn't enough to make me not dread collecting eggs. I hated that rooster.

So, yes, I suppose there was a time when our eggs were pretty fresh. But, it's not something that my mom was ever particularly worried about. She has never been finicky with the ingredients that she uses. But, she always manages to make something wonderful with whatever she has in the house. And Mom's scrambled eggs are no exception. As evidence of her easy approach, she finishes her eggs by melting a little processed American cheese on top of them. Absolutely nothing fancy about Kraft cheese slices! But, they are always perfect. The cheese adds a surprisingly sweet creaminess that I have always adored. I don't think they'd end up as good if she used anything else. Mom's recipe and ingredients probably wouldn't be mistaken for those at a Michelin starred restaurant, but that doesn't mean their scrambled eggs are better than my mom's.

So, what makes her scrambled eggs so good? Without fail, they are flavorful and light; they have a soft and airy texture without being runny. I must admit that I don't know how she pulls this off. I mean, whenever she makes eggs, she doesn't exactly pay close attention to what she's doing or watch how the eggs are coming along as she cooks them. For her, egg making is usually accompanied by something else. It could be playing with her grandchildren, frying bacon, or washing and cutting fruit. And, truth be told, I think she'd be the first to say that she's not the most observant while she's making them either. If asked how she did it, she'd probably say, "Oh, I don't know. Mix some eggs with some salt and milk and then cook them until they're done." (This is her stock response for how she cooks everything. "Add a little of this and a little of that. I don't really pay too much attention to amounts anymore.")

TREAT EGGS WITH THE RESPECT THEY DESERVE

Entry into the world of being a chef at a fine dining establishment is typically an apprenticed trade. Aspiring chefs often start off cleaning dishes before moving on to prepping ingredients and then to working at the different stations in the kitchen. Even if you attend a culinary school, you are often expected to pay your dues in some way before you get a job with any cachet. So, there is no surprise that interviews for these jobs often include demonstrations of skill in the kitchen.

Certainly chefs are known for working under pressure. The time and attention demands of working in a restaurant during a dinner rush can be more than a little daunting. This pressure is heightened by the expectation of perfection. Even though a chef might be used to working in this tense environment, I can't imagine the anxiety of performing, in an interview, in front of a head chef. When that chef is famous, or famously demanding, I don't know how applicants can distinguish a pot from a pan.

When chefs tell stories about their experiences, there is one interview task that comes up time and time again, making them go into anxiety-riddled flashbacks. Making an omelet. They are not asked to reproduce a dish from a head chef's recipe. They are not asked to make a dish from one of their own recipes. They are asked to make an omelet. One ingredient. One pan. And a little butter and salt. There is no hiding or covering up poor technique or any chance for bonus points. Either you can do it or you can't. For all of the creativity and skill that chefs gain in their apprenticeships, the omelet is proof that a chef pays attention to detail and is skilled at reading cues from their dishes.

In a wonderful series of interviews, *bon appétit* magazine asked several chefs about their first perfect omelet. There are enlightening and colorful answers to this question, including a statement from Wolfgang Puck on what he thought of the omelet-making capabilities of several contestants on a reality-TV cooking show.[1] But, I think that my favorite answer came from José Andrés, one of the most playful chefs in the world, whose restaurant network is based out of Washington, DC. "You have to feel the egg in order to know exactly when it is ready. A recipe you can give away, but that feeling you cannot teach. It's not an easy task. You have to be gentle with the pan, control the temperature, don't stir too much or too little, and make sure you have just enough oil or butter."[1]

There is something about that, isn't there? Even I can follow a complicated recipe. But, just because I can follow it, doesn't mean that the resulting meal will be any good. Perfection, whatever your definition, only comes from attention and understanding and adapting. The same is true in any chemistry lab. Just because a protocol has been peer reviewed and published doesn't mean that any shmuck with a round bottom flask will be able to make it work. Often there is an ineffable observation that corresponds to getting a procedure right. The ability to make this observation at the right time and, then, to respond correctly, only comes from the experience of having tried the experiment, protocol, recipe a hundred times.

In testing an aspiring chef, omelet making is meant to show two things. One, a perfect dish can come from something as simple and unassuming as an egg. Two, arriving at perfection is only possible when a chef pays close attention to their ingredients, understands how they need to be treated, and transforms them through cooking.

Certainly, there are multiple ways of arriving at perfection, whatever that may be. But, each path requires the right combination of experience and deft touch. As an example of this, Chef Gordon Ramsay's scrambled eggs are entirely different from my mom's.

Chef Ramsay's scrambled egg-making skill has been well documented in interviews and videos.[2] To him, the perfect scrambled eggs are more like a just barely thickened and set egg sauce. They contain no cream and are salted only after they have been cooked. He whips his eggs in a bowl, places them in a pot, cooks them for a minute, removes them from the heat, stirs them like mad with a fork or whisk, and repeats the heating and mixing steps until the eggs are just starting to firm up. He has a vision. He understands how to get from the whipped eggs to his final product. And, he understands how to respond to the changes the eggs are going through.

Chef Ramsay's recipe may require more work and immediate attention than my mother's scrambled eggs. However, that doesn't mean that my mom's requires any less understanding or expertise. Part of what makes her scrambled eggs so good (and why she can multitask while making them) is that she has made them a thousand times. She knows exactly what temperature the pan needs to be set at. She knows exactly when and how often they need to be stirred. And she knows exactly when they'll be done. She understands perfectly what she can do with her time while she's making scrambled eggs.

INCREDIBLE, EDIBLE SCIENCE

Dishes that have eggs as their primary ingredient, like scrambled eggs and omelets, can illustrate how eggs are used in more complex recipes: ice creams, crème brûlées, meringues, soufflés, and others. In all of these foods, you start off with a liquid mixture. As it cooks, the liquid thickens and becomes more viscous. How thick it ends up is entirely dependent upon how long you cook it and the temperature at which you cook it. Pretty generic. Pretty standard. However, understanding what happens as you cook an egg can give you a lot of control over how well your finished dish ends up.

For all of these foods, the part of the egg that dictates texture is primarily the group of proteins in the yolk and the egg white.

Proteins are molecular machines that organize and run life on earth.

By most accounts, we give credit for our life to our genetic material, DNA. Now, DNA is a truly incredible substance. The structure and makeup of DNA has been honed by evolution to withstand and rebound from all sorts of damage caused by living. It can wind itself up into compact structures. (If you were to uncoil DNA into a straight line, it would stretch out to somewhere between 1 and 2 meters. And, somehow, all of that DNA is able to squeeze itself inside the nucleus of a cell, which is just under 6 micrometers, or 6 millionths of a meter, in diameter.[3]) Most importantly, though, DNA is a blueprint for how to make proteins. The sequence of DNA gives a description of how big a protein should be, how the protein should be put together, and, often, what part of a cell a growing protein should be placed in.

If DNA tells us what life is, then proteins tell us how we live our lives. Proteins make us move. They give us taste and smell and sight and touch and hearing. Proteins give us identity. As an extreme example of this, consider identical twins. Even though identical twins have the same set of DNA inside of their cells, the differences in how their proteins develop make them unique individuals. In the old nature *versus* nurture debate surrounding who we become, the example of identical twins shows us the types of effects that nurturing can have on developing our proteins and our identities.

You can think of a protein like a train. Just as trains are made of multiple cars, linked together end-to-end, proteins are chains of amino acids. And just like train cars contain different cargo (automobiles, livestock, crops, industrial equipment, *etc.*), there are different kinds of amino acids that can go into making a protein. Humans have twenty amino acids that constitute our proteins. There are several others that do get used in our proteins on a rare basis, but these twenty are what evolution has decided we need most. In our proteins, unlike on a train, the amino acids aren't necessarily sorted by type. That is, all of the cattle cars don't necessarily have to be next to one another.

PROTEINS

Protein Architecture

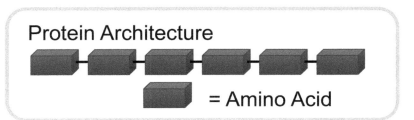

= Amino Acid

Amino Acids

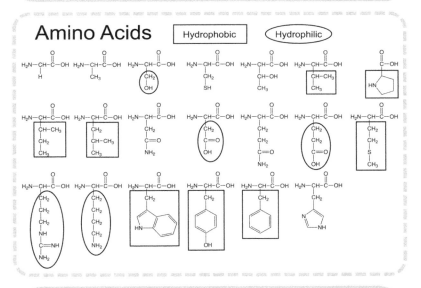

Protein Shapes

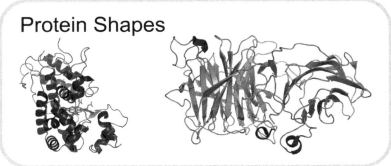

Protein variety and identity is partly defined by this evolutionary defined ordering of amino acids. Proteins are also defined by their length, which can vary between the tens and thousands of amino acids. The sequence of the protein is what ultimately gives rise to the job or task that protein carries out in our everyday lives.

One of the most important properties of any protein is its shape. This is where my train analogy fails miserably. A train's cars are held in a straight line by the tracks that they are sitting on. A protein, however, does not have these constraints. And, a protein is not content to just hang out, all unraveled, and have all of its amino acids interacting with the water inside of our bodies. There are many types of chemical and physical interaction that give rise to protein shape, but we can generically discuss the major drivers of their shape as being hydrophobic and hydrophilic forces.

The inside of a cell is, for the most part, a watery environment. A protein has to reside and function within that environment. Looking at the amino acids individually: some of them have positive charges and want to interact with a negative charge; some of them are hydrophobic and want to interact with other hydrophobic, or oily, things; some of them are hydrophilic and want to interact with water; and some amino acids aren't strongly hydrophobic or hydrophilic and don't strongly drive interactions one way or the other. When you take a full protein, which is made up of a seemingly random arrangement of all of these types of amino acids, and put it into water, the protein is going to change its shape so that the hydrophobic amino acids are going to interact with other hydrophobic amino acids and that the hydrophilic amino acids are going to interact with water. In general, this means that the hydrophobic amino acids are found in the middle of a crumpled protein structure and the hydrophilic amino acids are found on the outside of that structure. While this makes intuitive sense, getting a protein to take on an optimal structure requires a complicated number of origami-like folding steps.

Even though this protein origami can result in a nearly infinite number of forms, there are several shapes that occur over and over. There are individual strands that fold back on themselves multiple times to form sheets. There are spring-like helices. And, there are parts of the protein that remain unstructured.

The final geometry that a protein takes on is often its most stable shape. In chemistry terms, this shape represents its lowest energy structure. A protein is not necessarily content to just hang out in that shape though. Natural energy from the environment

Eggs

(temperature, light, motion) makes the protein wiggle and jiggle. Although this jitteriness removes the protein from its most stable form, jiggling is an essential requirement for the protein to do the tasks they are supposed to do in our bodies.[4] An extension of this normal flexibility is that, if a protein takes on too much energy, the small motions become big motions and the protein shape can change drastically. One of the important things that happens in these changes is that the hydrophobic amino acids, which had been tucked away inside of the folded protein, can be turned out and be forced to interact with water.

Just like in coffee flavor extraction (discussed in Chapter 1), the hydrophobic amino acids have no (or very few) stabilizing interactions with water. When proteins experience extra added energy (increased amounts of temperature, light, or motion) that energy compensates for the weak interactions. What this means is that excess energy can take the intricately formed protein-origami and unravel it, leaving almost no trace of the shape it once had.

Another way to (very simplistically) imagine what is going on here is to think of a single Post-It Note sheet. The Post-it Note is in its most stable shape whenever the sheet is folded over in a way that the adhesive is touching another part of the paper. When it's in this shape, the paper doesn't need to have any other interactions. That is, if you try to put it on a wall, it will fall down to the ground. However, if you give the system enough energy (by pulling on the paper) you can unfold it. And when you unfold it, it becomes sticky again. The part with the glue on it has become exposed, even though it would prefer to be stuck to something else.

EGG-CELLENT RESEARCH

César Vega (an acquaintance of mine) is a food scientist who works for an industrial food manufacturing company. Not only is César an excellent scientist, he received his PhD in food science from University College Cork, but he is also a classically trained chef (having studied at *Le Cordon Bleu*), and an amazing communicator. I have seen him give a keynote lecture at a conference in front of a crowd that pushed towards a thousand people. While his topic, ice cream chemistry, was certainly responsible for the size of that crowd, his talk was brilliant. He was engaging and entertaining and the audience, which was made up of professional chemists, went away with a better appreciation and understanding of how chemistry works in food production. To give you an idea of how anticipated his talk was, I gave a lecture at the same conference, which boasted an attendance of roughly 50 people.

César is also a very capable writer. He edited a book titled *The Kitchen as Laboratory*,[5] which I highly suggest reading. In this book, Cesar and several other authors (people who have been trained as either scientists or chefs or both) explore the ways they use science in their personal and professional kitchens, to better understand and improve the food that they cook. It would certainly be appropriate to say that *The Kitchen as Laboratory* has informed, and was an inspiration for, the book you are currently reading.

One of the primary recurring topics in that book is modernist cuisine, also known as molecular gastronomy. While neither of these names does full justice to this genre of cooking, they do capture some aspect of its essence: namely, precision. In this style of cooking, different ingredients and new techniques are used to exercise control over the cooking process. Precision comes in the form of using ingredients that perform a very precise role, adding precise amounts of these ingredients, precisely controlling temperature and other cooking tasks, and having influence over the precise textures and aromas and appearance over the dishes that are cooked.

In César's chapter he describes cooking eggs using *sous vide* methods, in which a cook has very precise control over both temperature and time of food preparation.[6] Despite the French name, meaning under vacuum, which doesn't really indicate how this works, *sous vide* cooking involves placing food in a pot of water that has very fine temperature control. What makes *sous vide* cooking different from baking something in an oven or boiling in a pot of water or frying on a pan, is that the temperature of the water is set to the exact temperature you want your food to come to. For example, a steak that has been cooked to medium has an internal temperature of around 57 or 58 °C (135 or 136 °F). A grill or frying pan can reach temperatures over hundreds of degrees Celsius. To make a steak medium, a chef has to have the meat on the grill for just the right amount of time. Too little time, and the steak is rare. Too much time, and the steak is over-done. For *sous vide* cooking, the pot of water is set to 57 °C, making it close to impossible to overcook your food. For dishes like steak, you need to place the food in vacuum-sealed packages (the source of the name *sous vide*) so that you don't leach flavor compounds into the water. An egg doesn't need packaging as its shell acts as a convenient container.

César was interested in a preparation called a 6X °C egg. The X in this instance stands for another number. One could make a 63 °C egg or a 67 °C egg. Chefs had found that by changing the temperature (and the total time the egg was cooked), they could produce yolks with just the right amount of runniness. César was curious about this not only because the 6X °C egg was becoming so common in fine dining establishments, but also because there had never been any full scientific studies done on how time and temperature affect the consistency of a cooked egg yolk. So César did this study himself, publishing his work in the science journal, *Food Biophysics*,[7] and describing the work for a general audience in *The Kitchen as Laboratory*.

Now, when you heat an egg, the proteins in the egg unravel. Temperature is a form of energy, and extra, added energy can cause a protein to wiggle so much that it loses its shape, causing its hydrophobic amino acids to stick out into the water. Temperature (energy) has two effects on the unfolding of proteins. First, the more energy you add, the faster a protein unfolds. Imagine a ball of string. That string could be unraveled pretty quickly if pulled apart by two automobiles. These cars would straighten the string much faster than I could pulling with my hands, precisely because the cars could exert more energy. Second, the amount, or extent of protein unfolding, is also dependent upon the amount of energy that you put in. Say the ball of string has a snag in it. If I am not strong enough to undo that snag, then I won't ever be able to unravel the string completely. The two cars, pulling in opposite directions, would likely have enough energy to do this. So, the higher the temperature, the faster and more complete the protein unfolding will be.

The texture of cooked egg is a result of what happens to the proteins after they unfold. Going back to the Post-it Note analogy. Imagine that you have two Post-its in which are lightly folded (*i.e.* not creased) down the middle, such that the part with the glue is stuck to itself and the non-sticky part is curved. If you completely unfold them, the two Post-its can lay flat together, stuck on one another. Next consider only partly unfolding the Post-its, so that just the corners are exposed. These corners are sticky and can be attached to another Post-it. This latter combination of

Post-its retains some springiness and lightness in its structure, even though the two sheets are attached to one another. In the former scenario, this texture is completely lost, as the two sheets just lay flat on one another. When cooking eggs, this unfolding and aggregation happens millions and millions of times as the network of interacting proteins grows larger and more complex. By controlling the amount of protein unfolding, you can control the network that you form and, ultimately, the texture of the eggs that you cook. Another name for this type of network (polymers that form an interconnected web) is hydrogel, called so because of the ability of these networks to trap water molecules. I will refer to hydrogels multiple times in this book because of their relevance to cooking, whether they are made by egg proteins or some other food polymer.

It was no mistake that Vega chose to study egg yolks in his research. For many culinary dishes, the success of preparation hinges upon the precise texture of the cooked yolk. In some dishes, you want the yolk to act as a sauce that coats the other food on the plate. In other dishes, you want the yolk to become more solid with, say, the consistency of Nutella, so that it is experienced more

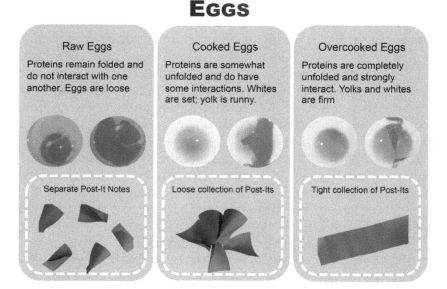

Eggs

on its own. This reasoning should be fairly obvious to most cooks. The inspired choice of studying only egg yolks, though, is that when uncooked eggs are mixed with other ingredients, the egg yolks tend to dominate how the eggs control the final texture of the food. This happens because the yolks contain molecules called emulsifiers. These molecules are interesting in their own right and we will explore them further in the chapter on vinaigrettes. What is important for now, is that emulsifiers help hydrophobic and hydrophilic things to mix evenly. These molecules greatly affect the aggregation and final consistency of cooked egg yolks. Understanding yolk texture as a function of cooking, therefore, is a stand-in for understanding how custards (mixtures of egg yolks and cream), soufflés, and batters change in texture as they cook.

For his research, Vega measured the amount of time required, cooking at one temperature, for egg yolks to reach a particular viscosity. Viscosity is just a measure of how a liquid resists deformation when it is exposed to some stress. In English, viscosity is the property that tells you how hard you have to chew your food so that it becomes squishy. When we're talking about liquids, their viscosities vary by small amounts. But, intuitively, we understand the difference of viscosities between water and whipped cream and peanut butter. Performing this measurement at multiple temperatures, Vega was able to produce a database of time required for an egg yolk to reach a desired consistency (sour cream, mayonnaise, honey, or toothpaste, to name but a few) at a specific temperature. For instance, to reach the consistency of honey, an egg yolk must be cooked at 61 °C for over 150 minutes or at 64 °C for roughly 25 minutes. As you might expect, he also found that a chef's margin of error for producing a specific consistency decreases at higher temperatures. That is, at 68 °C, it only takes a little over 10 minutes for an egg yolk to go from the viscosity of condensed milk to the viscosity of Marmite. At 63 °C this change in viscosity occurs over around 100 minutes. Vega also found that for temperatures below 60 °C, the amount of time required to observe any useful change of viscosity was much longer than any person would ever have the patience for when making eggs at home.

Vega's research provides a really useful guide for professional and home chefs to get the yolk consistency that they demand for their dishes. His research also makes clear, to food scientists and other food science-curious individuals, in the more technical details of cooking eggs.[7] Diving deeper into the protein chemistry, he found that the proteins α-livetin and avian immunoglobulin Y, are mostly responsible for the texture formation processes in cooked egg yolks, while several other proteins (the low-density lipoproteins) play a very minor role.

CHEF RAMSAY *VS*. MY MOM

So what can we say about the way that Chef Ramsay cooks his scrambled eggs in comparison to the way my mom does. Aside from the fact that my mom would win, which should have been more than obvious by now, can we use Vega's (and others) research to understand and make the kind of eggs we want?

Well, of course!

But, before we get into that, let's talk about really bad scrambled eggs. These come in two distinct varieties: runny and dry.

Runny eggs are made when they are heated unevenly. Some parts of the eggs get cooked while others do not. This can be caused by some combination of using too narrow of a pan and not enough stirring. The result that you get in the end includes clumps of solid eggs surrounded by a thin runny liquid.

Dry eggs have been overcooked to the point where they become desiccated and chewy. Thinking back to our Post-it notes, the hydrogel that gets formed by the partially unfolded proteins contains space within the protein network for water to hang out. For the case of the fully unfolded proteins, the hydrogel network has basically collapsed on itself and no longer contains any room for water. Dry eggs are made when we overcook the eggs, completely unfolding the proteins, and resulting in a hydrogel that does not contain any water. That is, dry eggs are dry because you've squeezed all of the water out of them. Dry eggs also tend to taste rubbery. Part of the reason for this is that you actually start to create sulfur-containing flavor compounds (from sulfur-containing amino acids found in the egg proteins) when you cook eggs too long.

Unfortunately, it is easy to overcook eggs. The pan that you are cooking with on the stove can easily reach temperatures topping several hundred degrees Celsius. As Vega proved with his research on yolk consistency, the window of time that you have between underdone, perfectly done, and over-done, gets smaller

Eggs

and smaller as you increase the temperature. A moment of inattention can see your eggs go from perfection to the trash. The easiest way to prevent this is to just turn down the heat on your pan; make your window for error a whole lot bigger.

One of the main differences between how Chef Ramsay approaches making scrambled eggs and the way my mom approaches this, is in their use of salt. Chef Ramsay salts his eggs only after they are finished cooking. My mom salts her eggs before. Harold McGee gives some correct, but misleading and seemingly contradictory, advice in *On Food and Cooking*.[8] In one place, he writes that salt can cause the egg proteins to aggregate before they are cooked and can lead to a drier dish. In another place, he writes that salt can be used to shield positive and negative charges on the unfolded egg proteins, which can keep egg proteins from over-aggregating and drying out. So, which is it? The answer is that it's a little bit of both. And, knowing when to salt, depending upon how you're cooking your eggs, is the most important part.

When you dissolve salt (typically sodium chloride) in water, the solid chunks of salt break down into lots of sodium ions (which have a positive charge) and lots of chloride ions (which have a negative charge). Normally, positively charged things are attracted to negatively charged things and will stick together. (The corollary of this is that negative charges repel each other, and positive charges repel each other.) However, water is also attracted to the charges. And, when there is enough water around the positive and negative charges, they don't mind so much that they are not attached to one another.

Many egg proteins have an overall negative charge, due to the number of negatively charged amino acids that they contain. In their fully folded state, the proteins are compact. Because of this, there is a large charge density (lots of negative charge) spread over the small surface of the folded protein. The proteins in eggs then repel one another because of these charges. If you add enough salt to an uncooked egg, the positive sodium ions will interact with the negative charges on the protein. The overall effect of this is that the net charge on the protein goes away. When the proteins become neutral (not positive or negative) they will stop repelling one another and, while still folded, will aggregate. Increased salt (and sugar) concentrations can, in some circumstances, also help in stabilizing a protein structure. These increased amounts of salt result in making it more difficult to unfold an egg protein, causing a chef to have to add more energy to do so. Therefore, adding salt before unfolding egg proteins can be detrimental to the final texture of the eggs.

On the other hand, in an unfolded protein, the charges are spread out along the entire elongated polymer. While the protein still has a negative overall charge, there are some positive charges as well. As the unfolded protein aggregates to make the hydrogel, the primary driving force behind this aggregation is the hydrophobic–hydrophobic attraction. However, positively charged amino acids interacting with negatively charged amino acids can also strengthen the hydrogel. As we discussed earlier, when the interactions between the proteins become too strong, water gets squeezed out and we are left with dry and crumbly eggs. When the right amount of salt is added, the positive and negative charges on the protein are taken up by chloride and sodium ions, respectively. Therefore, after the proteins have been unfolded, but before the eggs have been fully heated, added salt can help the eggs to stay tender as they are cooked.

CHARGE INTERACTIONS

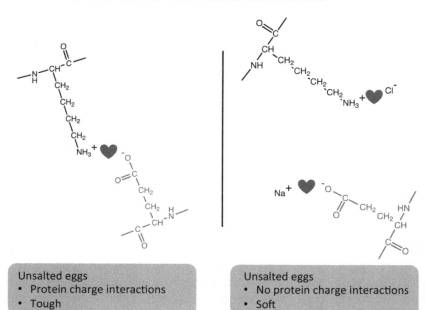

Unsalted eggs
- Protein charge interactions
- Tough

Unsalted eggs
- No protein charge interactions
- Soft

When Chef Ramsay cooks his eggs, he is basically doing a less controlled version of Cesar Vega's experiments: slow heating over a long period of time to get the desired consistency he's looking for. The difference between the two is the speed of heat transfer to the egg. In Vega's experiments, there is a delay time while the temperature of the egg increases. For Chef Ramsay, the egg is spread out across the width of the pan and can be heated to the proper temperature pretty quickly. To make sure that he doesn't overheat his egg proteins, he removes the pan from the heat and stirs the eggs to make sure they are heated slowly and evenly. The other way he does this is by making sure that the burner on his pan is not turned up too high. Because he wants his scrambled eggs to have the consistency of a custard, he has to make sure that none of the proteins over aggregate. That aggregation is, in part, prevented by not adding salt until he is finished. Because his proteins unfold and aggregate so slowly, the added salt would cause the folded proteins to aggregate before they are cooked. Along with this, keeping his temperature low allows Chef Ramsay to increase the window of time in which to make sure he doesn't overcook his eggs.

My mom wants a different consistency to her scrambled eggs. Hers go past the point of being a custard and become a firmer semisolid. There are a couple tricks that she uses to keep them tender as they cook.

The first thing that she does is whip the eggs in a bowl with a little cream. This has two effects. First, whipping the eggs starts to unfold the uncooked proteins. Just as the added heat (in the form of a hot pan) can give the proteins the energy they need to unfold, the mechanical energy that comes from whipping can also cause the proteins to change shape. As the proteins unfold, they start to aggregate, and they do so around the air bubbles that are being incorporated by whipping. This can be seen more dramatically when whipping egg whites on their own. The whites, once a puddle of liquid, quickly become a foam. As the whipping continues, the foam becomes stiffer (caused by the proteins becoming more aggregated). This happens in the whole egg-cream mixture as well. And, while bubbles are incorporated, the aggregation of the proteins around those bubbles is tempered by the fats and emulsifiers in the yolk. (The ability of the fats from the yolk to diminish the strength of the foam is the reason why you cannot have any amount of yolk fall into your egg whites when you are making meringues.) For the scrambled eggs, the bubbles translate to a softer, lighter texture of the final dish.

The cream also adds a certain, well, creaminess to the eggs. The extra fats from the milk become part of the hydrogel network, trapped within the unfolded and aggregated proteins. The milk also helps to spread the proteins out. Just because you dilute the egg proteins does not mean that you cannot form the protein mesh. We can look at the recipes from some other egg-cream mixtures to see how spreading the proteins out doesn't prevent hydrogel formation. In scrambled eggs, there is usually just a splash of milk mixed in. For ice cream, you use 675 milliliters (mL) (about 3 cups) of milk with 5 or 6 egg yolks. For flan, it's around 750 mL (just over 3 cups) of cream with 3 egg yolks. And, for crème brûlée, it's over 900 mL (almost 4 cups) of cream with 6 egg yolks. In all of these other (not scrambled egg) recipes, diluting the concentration of egg proteins does not remove the ability of the egg proteins to form a hydrogel network. However, the overall effect of diluting the eggs is that it requires a lot of extra cooking for that network to over-aggregate and become dry and crumbly.

Once the eggs and cream have been whipped, it's time to add a little salt to help prevent over-aggregation. At this point, my mom's approach starts to mimic Vega's and Ramsay's. She cooks those eggs over a very low heat until she gets the consistency that she's looking for. That expanded window of time gives her the ability to play with the kids, make bacon, wash fruit, or just relax, because it's the weekend and she deserves it.

Now, I will permit Chef Ramsay and his family to disagree with my assessment of his scrambled eggs in comparison to my mom's. As for the rest of you ... you've got no excuse.

REFERENCES

1. M. Y. Park, *My First Perfect Omelet: Great Chefs Remember* [document on the internet]. bon appétit; 2014 October 14 [cited 2016 July 5], available from: http://www.bonappetit.com/people/chefs/article/perfect-omelet-chefs.

2. F. Cloake, *How to Make Perfect Scrambled Eggs* [document on the internet]. The Guardian; 2010 November 11 [cited 2016 July 5], available from: https://www.theguardian.com/lifeandstyle/wordofmouth/2010/nov/11/how-cook-perfect-scrambled-eggs.
3. B. Alberts, A. Johnson, J. Lewis, D. Morgan, M. Raff, K. Roberts and P. Walter, *Molecular Biology of the Cell*, Garland Science, New York, 2002.
4. E. Z. Eisenmesser, O. Millet, W. Labeikovsky, D. M. Korzhnev, M. Wolf-Watz, D. A. Bosco, J. J. Skalicky, L. E. Kay and D. Kern, Intrinsic Dynamics of an Enzyme Underlies Catalysis, *Nature*, 2005, **438**, 117–121.
5. *The Kitchen as Laboratory*, ed. C. Vega, J. Ubbink and E. van der Linden, Columbia University Press, New York, 2012.
6. C. Vega, Egg Yolk: A Library of Textures, in *The Kitchen as Laboratory*, ed. C. Vega, J. Ubbink and E. van der Linden, Columbia University Press, New York, 2012, pp. 134–141.
7. C. Vega and R. Mercadé-Prieto, culinary Biophysics: On the Nature of the 6X°C Egg, *Food Biophys.*, 2011, **6**, 152–159.
8. H. McGee, *On Food and Cooking*, Scribner, New York City, 2004.

CHAPTER 4

Pancakes

Being a dad, it is incumbent upon me to make breakfast on Saturday mornings. The workweeks are always long. Projects, deadlines, colleagues, unresponsive colleagues, commutes, and parenting. And, did I mention parenting. Time goes by quickly, but the days are always so long.

The magic of a lazy Saturday morning cannot be denied. There is nothing to think about except what's on your mind. Nobody has to rush to get dressed or finish forgotten homework or read and reply to "urgent" emails from colleagues. Being a parent, one of my biggest responsibilities in life is to facilitate freedom. I spend all week long policing freedom, pulling people (including myself) back to reality and the necessities of the schedule. Perhaps Saturday morning is my penance for the rest of the week.

When my kids were younger, Saturday mornings were dictated by their schedules. Get me out of my crib! Change my diaper! Give me food! Erika and I got into the routine of trading weekend mornings so that the other could have a little peace. Now that they are all past the diapering and playing with electrical sockets phase of their youth, they get their own time. Jack is five.

Chemistry in Your Kitchen
By Matthew Hartings
© Matthew Hartings, 2017
Published by the Royal Society of Chemistry, www.rsc.org

His curiosity and short attention span lead him on a random walk through his Saturday mornings. Within a 15 minute time span you will find him playing a video game, coloring, playing a different video game, hovering over his big sisters, and bouncing a basketball through the house. Claire is our quiet, self-absorbed-within-her-own-world child. She has her own secret adventures through the basement or in the yard or with her toys. She's happy as long as she's making the rules, which, to a middle child, doesn't feel like it comes all that often. Kaitlyn is our oldest. Her perfect Saturday morning involves sitting on the couch with the television remote in her hand finding just the right episode of just the right show. Although she probably complains the least of our children about her weekly schedule, she is certainly more put-upon by her packed schedule than our other kids. A little veg time on the sofa is entirely appropriate for this busy kid.

Erika deserves her own time and space just as much, if not more, than our children. As she is the official "keeper of the schedule" she takes on the bulk of the responsibility for our kids' daily activities, on top of which is her own busy freelancing work.

There is a real sense of both duty and pleasure in making breakfast on these mornings.

Like my own dad, there are a lot of times when I find donuts to be the best option. In the town where I grew up (Piqua, Ohio), there is a bakery housed inside of an unassuming grocery store named Ulbrich's. Our little town is not known for too many things. We used to have an underwear festival celebrating the long underwear factory located there. Located near Dayton, Ohio, home of the Wright Brothers, we do have some historical aviation-based industry. (I always try to find Hartzel propellers when looking at small airplanes.) Our town's quirkiness seems to come out in the curious slogans we've had over the years: "Centered in a World of Opportunity" and, my favorite, "Where Vision Becomes Reality." What we should be known for, though, are the donuts at Ulbrich's.

Here in Maryland, we don't have anything nearly as good as Ulbrich's (or as good as the blueberry donuts from Erika's hometown at White House Fruit Farm). But we do have a Krispy Kreme. Our kids love going there, probably because they get to watch the donuts being made on an ever-moving conveyer belt. The donuts are proofed, fried, flipped, fried some more, glazed with sugar, and cooled. They are a little piece of heaven hot off the belt. But, even when we they don't want to go to the store for the theatrics, that's where the kids want me to get donuts from.

Because the best ones are spent bumming around the house, Saturday mornings most often include making breakfast.

There is very little agreement in our household at mealtime. The two meals that everyone seems to get behind are bacon, lettuce, and tomato sandwiches and pizza. Other than that, there are complaints and arguments. Saturday morning arguments, usually revolve around whether to make pancakes or waffles.

Making good pancakes was the first enjoyable badge of honor that I earned as a father. The first badges were probably earned for taking some nighttime feedings and changing diapers. (Erika's virtual trophy cases rightly mock my badges.) Pancakes seemed like a very dad thing to do. They were the perfect consistency for when the kids were getting used to solid foods. They are reminiscent of my own happy childhood. And, for many reasons, they have a manly-responsibility tinged connoted association for me.

I have many reasons for taking this seemingly chauvinist point of view. Pancakes just make me think of dads. Maybe they're simple enough to make that even a dad can handle it. Maybe it's because every fundraiser I attended as a child (for some local civic group like the Kiwanis or Rotary Clubs) was a pancake breakfast where older men were running the skillets.

I may be conditioned to notice the dad-pancake connection. But, I see it in Erika's family as well. She tells stories about how her great grandfather would fry her a pancake in a cast iron pan filled with bacon grease. My own children expect their grandfather, Erika's dad, to make blueberry pancakes when we are visiting.

Between my mom and dad, Mom has probably made more pancakes for us kids than Dad has. However, of the two, my dad is the more adventurous one in his cooking. (Mom likes to go with what she knows and doesn't use recipes as much as memory. At least, that's what I notice.) Dad is always on the

Pancakes

lookout for some new out-there ingredient or recipe that he can use to impress us. This eternal search has found its way into his pancake breakfasts, too. I recall a recent visit to Mom and Dad where Dad showed off a pancake made with ricotta and lemon curd that he was especially proud of. (They were excellent, by the way.)

In some ways, I am a perfect mix of my parent's adventurousness. I love trying new recipes and seeing how other people make different dishes. But, if I use a new recipe, it had better be better than the recipe I am used to!

Telling then, is the anecdote that I have only used one pancake recipe for the past 9 years. Deborah Madison's cookbook *Vegetarian Cooking for Everyone* is a modern classic and includes some really nice southwestern dishes.[1] When we got this book, I had expected to use it most for its vegetarian main courses and as a place to get different ideas for side dishes. Surprisingly, the section on breakfast foods has been my favorite. I have made her pancake recipe so often that I can't remember the last time I looked in the book to look up the ingredients.

Mix the dry ingredients [192 grams (1 1/2 cups) of flour, 8 grams (2 teaspoons) of baking powder, 4 grams (1 teaspoon) of baking soda, 45 grams (3 tablespoons) of sugar, and salt to taste]. Mix the wet ingredients [367 grams (1 1/2 cups of buttermilk), 2 eggs, 42 grams (3 tablespoons) of melted butter, and 5 mL (1 teaspoon) of vanilla]. Combine the wet ingredients and the dry ingredients. Ladle into a hot pan (no oil or butter added) and cook.

The first several times I made this recipe, I wasn't so sure about it. It was good. But there was something a little off about it. A bitterness in the back of my mouth would show up as I was finishing a bite. I really wanted to like this recipe.

The third time I made it, I asked Erika what she thought about it. She thought it was fine, but that there was something funny about it. I went back to look at the recipe to make sure I was doing things right. (It is amazing how often chemists actually try to run a reaction – or a recipe – without actually thinking about the starting materials or analyzing the process a little more deeply.)

I had followed the directions to a T. Well, except for that I had used milk instead of buttermilk. Buttermilk, at the time, wasn't something that we kept in our house. For whatever reason I hadn't thought that it would be an issue. As I read through the recipe again, the milk-buttermilk thing didn't stand out right away. What stood out was the baking soda. Baking soda is one of the few bases we use in our cooking. (The excellent cookbook *Bitter* describes, in wonderful detail, many more.[2]) Most of the foods we eat are acidic. Acids tend to taste sour. Bases taste bitter. So, I found the source of strange bitterness from the recipe. Now I just had to figure out why it would be there in the first place.

Both baking soda and baking powder are chemical leaveners; they add gas to different baked goods to make them rise. Baking powder is self-sufficient. It contains sodium bicarbonate (a base), some sort of acid (different brands use different food-safe acids), and some starch to separate the acid and the base, preventing them from reacting with one another. Once they are mixed in water, however, the acid molecules and base molecules will bump into one another to react. This is the same principle by which Alka-Seltzer works. Drop a tablet into some water. The water dissolves the tablet. The molecules that had been trapped in solid form in the tablet are now free to move around in the liquid. They find each other. They react. The reaction produces carbon dioxide bubbles.

Baking soda is different from baking powder. Baking soda is pure sodium bicarbonate. It won't react on its own. It needs an acid to help make carbon dioxide for leavening.

"It needs an acid to make carbon dioxide. There's no acid in my recipe. I'm eating straight sodium bicarbonate in my pancakes. I am an idiot.", I thought. I was ready to revoke both my dad-card and my chemist-card. It was a low point for me on many levels. Science is often humbling. Hard work in the lab can yield lackluster experimental results. Research can be demanding. But, the most humbling times are when science reminds you of things you've already learned that you should have had the sense to remember.

I knew buttermilk was sour. I knew it was more acidic than milk. But I hadn't thought about the recipe at all.

Pancakes

ACIDS, BASES, pH

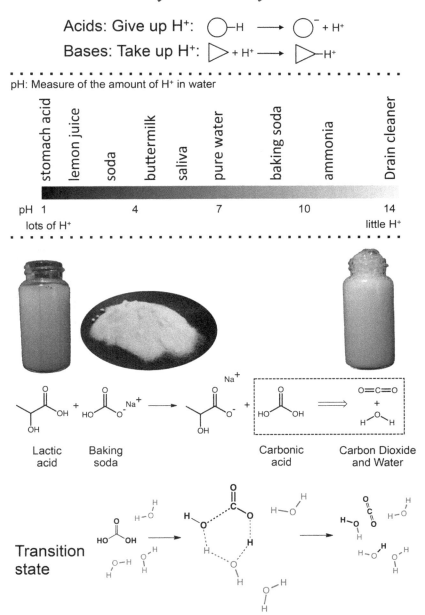

One of the skills I try to instill into my chemistry students is that they have to analyze any reaction before they try it. What are the starting materials? What is the mechanism? Does it generate heat? Are there any risks and how do we minimize them? How do we isolate what we're trying to make? A simple analysis will make the chemistry go better and will keep you safe in lab. I espouse this dogma so often that I require it of the students who take my chemistry of cooking class as well. Figure out why an ingredient is used, and then you can determine how much or how little of that ingredient you can get away with using in preparing a dish. You can also figure out how to substitute for that ingredient.

I had absolutely failed to live up to my own expectations as a chemist (and a father) by not analyzing this reaction (recipe).

I am sure that many of you are chuckling at my expense. The difference between baking soda and baking powder is instinctual to most of us who cook at home. The same is true for the difference between milk and buttermilk.

This reaction, in which bicarbonate is converted to a gas, is one of the most influential to both our culinary world and our environment, that it is nearly unforgivable that I hadn't properly considered it.

At its heart, the reaction between baking soda and buttermilk is just a simple acid base reaction. An acid is any molecule that gives up a positively charged hydrogen ion (H^+). (We chemists also refer to H^+, as a proton.) There are some molecules that you may remember from your days in chemistry lab as being acidic: hydrochloric acid, sulfuric acid, and acetic acid. The acid in buttermilk happens to be lactic acid. Add any of these acids to water, and they will give up an H^+ from their atomic architecture. Put hydrochloric acid (HCl) into water, it turns into H^+ and Cl^-. Put lactic acid ($CH_3CH(OH)COOH$) into water, it turns into H^+ and $CH_3CH(OH)COO^-$. You could literally put ANYTHING into water and, as long as ANYTHING gave up a proton (turning into H^+ and ANYTING$^-$), it would be considered an acid.

A base is any molecule that scavenges for any H^+ in solution, and adds it to its atomic framework. The prototypical laboratory acid is the hydroxide ion (OH^-). This ion will take on a proton

(H^+), resulting in the formation of water (HOH, or H_2O). Another common household base is ammonia (NH_3), which will take up protons to turn into ammonium (NH_4^+).

The pH scale is the way we discuss whether a solution (a watery mixture) is acidic or neutral or basic. If a solution at room temperature is acidic (pH values lower than 7) there are excess protons bouncing around in that solution. If a base is added to water, the result is that, instead of protons, excess hydroxide ions are present in water. (The hydroxide ions don't necessarily come from the base, though they can. They come from the way that the water interacts with the base.) Basic solutions at room temperature have high concentrations of OH^- and low concentrations of H^+ and pH values greater than 7.

When baking soda ($NaHCO_3$) and lactic acid ($CH_3CH(OH)COOH$) react, their identities as a base and an acid are the first things that manifest themselves. Lactic acid gives up a proton and baking soda takes that same proton for itself. The result of this initial reaction are two new compounds: sodium lactate ($CH_3CH(OH)COO^-Na^+$) and carbonic acid (H_2CO_3).

Carbonic acid (H_2CO_3), when placed in water, tends to break apart into water (H_2O) and carbon dioxide (CO_2). This is where all of those bubbles in your pancakes come from. Whether you use baking soda and an acidic ingredient (like buttermilk or vinegar or lemon juice) or baking powder, carbonic acid is formed and then decomposes into water and carbon dioxide.

This is such a vital chemical reaction for our lives on Earth that chemists have gone to great lengths to understand it. As a chemist, I find beauty in the ways that molecules arrange themselves. This is especially true for chemistries that are significant to our lives. The arrangement of carbonic acid molecules as they dissociate, is one of these. Any reaction is a chemical dance that finds the partners moving and twirling and vibrating until, at last, they separate from one another. It is in this separation that they find that they have changed. Molecules enter the dance with one identity and leave with an entirely different one. It is possible for molecules to find their way back to an old identity, but they will have to go through a different dance with different partners to arrive there.

In the middle of this molecular mambo, is a position that is the most difficult to hold. Each dancer must get themselves out of balance and find the right way to hold onto and embrace their partner. The tug of another can make it more difficult to situate yourself just right. Yet, there are moments of transcendence when it all comes together and the choreography yields a beauty that could never be achieved by a lone dancer.

This most-difficult-position in a chemical reaction is called a transition state. It is the position, having achieved it, chemicals will leave, changed from the way they entered. Many chemists obsess over transition states. The dance is beautiful. But the transition state lets us know just how difficult a reaction is, how long it will take, and how it might be made easier.

The molecular arrangement for the transition state in the decomposition of carbonic acid into water and carbon dioxide is no different.[3] The carbonic acid doesn't act on its own. There are other water molecules around that have to sit in the right place and tug on the carbonic acid in just the right way. In this dance, protons become entranced by the water and find themselves in a confused position in which they're not sure if they belong to the carbonic acid or the water. The confusion sorts itself out as one water molecule does steal a single proton, giving one of its own protons to an oxygen in the carbonic acid, which itself is removed, becoming a water molecule of its own, and leaving carbon dioxide.

Much like a set of dancers, chemicals can retrace their steps and move backwards to become what they once were. In chemical terms, these reactions manifest themselves in an equilibrium. A whole group of dancers that is continuously pushing forwards and backwards, always maintaining the same number of dancers in their original and altered positions.

This is a fact that we often don't teach our students in chemistry well enough. Our western societies are used to reading from left to right. And, when we write our reactions from left to right, our brains are conditioned to see a progression and finality in that direction.

All chemical reactions are capable of reversal, and most chemical reactions do find themselves in some sort of equilibrium as opposed to finishing after a single move from the left (starting materials) to the right (products). Even using those names, "starting materials" and "products", makes you think

Pancakes

that you start with one and end with the other and that's all there is to that. The decomposition of carbonic acid is one of these reactions that finds itself at equilibrium in our global environment. In fact, as we start adding more carbon dioxide into our atmosphere, we populate one side of the dance floor, and the reaction must adjust to populate more of the other side.

Stepping away from the dancing analogies and looking at our carbonic acid reaction, adding more carbon dioxide to water will result in the fabrication of larger amounts of carbonic acid.

This shifting equilibrium is a major concern for our environment. Carbonic acid is a tricky substance. When it is in water, it can be involved in two very important reactions. The first reaction we've been discussing. Through an equilibrium it bounces back and forth between carbonic acid and carbon dioxide. The second reaction sees carbonic acid acting like an actual acid, giving up its proton and turning into the bicarbonate ion. These reactions are connected, the carbonic

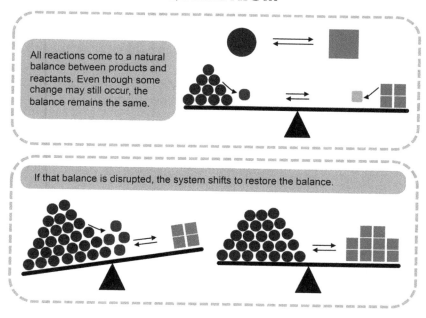

EQUILIBRIUM

All reactions come to a natural balance between products and reactants. Even though some change may still occur, the balance remains the same.

If that balance is disrupted, the system shifts to restore the balance.

acid bridging from carbon dioxide and water, to a proton and bicarbonate.

Through these relations, added carbon dioxide results in increased acidity.

The more carbon dioxide that gets put into water, the higher the acid content of that water will be.

The effects of carbon dioxide in the atmosphere on global warming are well documented. Carbon dioxide traps energy at the surface of the Earth. As more carbon dioxide is put into the atmosphere, more energy gets trapped and, on average, the temperature increases. There are numerous ecological effects of this increasing temperature. The first wave of changes has been seen in the oceans, not just with rising waters, but also manifesting in changes to living ecosystems. A 2016 report chronicled the mass bleaching of corals found in the Great Barrier Reef, putting them at risk of dying.[4] These reefs are home and shelter and food, for plants and animals and microbes. This type of bleaching is a predicted result of global warming and puts all of the life that depends upon the coral at risk.

There is another less discussed, but equally important, effect of increased carbon dioxide, and it relates to the equilibria of carbonic acid. As carbon dioxide is soaked up by the ocean, the pH of the ocean decreases. Higher ocean acidity can result in all sorts of destructive processes. Focusing again on coral. Coral beds are built upon the salt, calcium carbonate, which does not dissolve like common table salt. The fact that calcium carbonate is insoluble in water, is the basis for corals, mollusk shells, and a host of other oceanic materials that different organisms use for protection. At lower pH (higher acidity) the chemistry that maintains calcium carbonate as a solid changes. Calcium carbonate becomes ever more soluble in seawater. (To help to visualize this effect, imagine two pieces of chalk. Place one in regular seawater and another in low pH seawater. The chalk in regular seawater will maintain its size, while the one in low pH seawater will shrink to a smaller size.) In a world with lower pH oceans, the organisms who rely on calcium carbonate will have to continually expend extra energy to maintain the structures that they build.

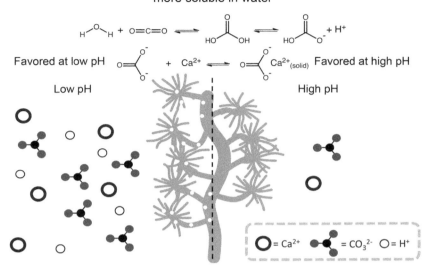

It is imperative that we start finding ways to both reduce our emission of carbon dioxide, and to extract the carbon dioxide in our atmosphere to maintain levels suitable for our life on Earth. Within the chemistry community, scientists of my generation have been trained to keep an eye on technologies that can mitigate atmospheric carbon dioxide. A general thrust of many research groups involves using CO_2 as a cheap and abundant natural feedstock to make more valuable and useful chemicals, things like polymers and medicines and materials. We call these value-added molecules, commodity chemicals. Much of this research takes the success of artificial fertilizer production, which uses nitrogen from the air for industrial processes, as an inspiration.

One such scientist who is hunting for ways to convert carbon dioxide into commodity chemicals is Alex Miller at the University of North Carolina in Chapel Hill. (Alex and I have been friends for many years.) Much like nitrogen, carbon dioxide is a stable molecule that does not react easily. In its dance with other molecules, it must become highly distorted from its original shape. Alex (and many others like him) are trying to make chemicals, called catalysts, that make these distorted geometries a little easier to come by. Thinking of carbon dioxide as one of our dancers, it has to bend backwards and keep its head near to, but without touching, the floor. This is nearly impossible to pull off on its own. But the catalysts that Alex makes are meant to support these distorted shapes allowing them to find balance without toppling. The trick that Alex is playing is that he isn't directly trying to turn carbon dioxide into commodity chemicals. He is trying to turn carbon dioxide into building blocks that are necessary to make commodity chemicals. Converting carbon dioxide into molecules that are more reactive, like methanol or formic acid or other small building-block molecules, would be a huge technological advance.

Of course Alex isn't alone in this hunt. The research fund that supports his work at North Carolina does so as part of their collaboration with researchers at 6 other universities. These researchers are just a small part of the much broader alliance of chemists and engineers and physicists and biologists and geologists, who are expanding the frontiers of science with the sole purpose of lowering the amount of carbon dioxide in the atmosphere.

Securing research funding is difficult, no matter where you are from. There is a political conflict with any science that focuses on global warming. The competing interests between industry and the environment have become entrenched. There is the argument that research on the environment is immediately seen as being anti-industry and *vice versa*. While the actual conflict is more fictitious than the political disagreements make it out to be, the perception is real. The research that Alex and similar-minded scientists are pursuing, seeks to benefit both parties in the political global warming conflict. The development of new chemistry may reduce global warming (environmental concern), but will only do so, if it can be deployed on an industrial level while generating economic benefit (industrial concern).

While this has become a common theme for the scientists of my generation, this is only the case because of the foresight and successful research of previous generations. Even I have dabbled in this kind of work, trying to make systems that employ biological molecules, such as proteins, to facilitate the conversion of carbon dioxide. One of the most influential papers that helped to set the framework for our thinking is titled, "Catalysis Research of Relevance to Carbon Management: Progress, Challenges, and Opportunities."[5] This scientific and technological overview, published in 2001, was authored by 35 top scientists and set the stage for countless experiments in labs all over the world.

Outside of this research, a number of scientists are looking for technologies that are advanced enough to more readily deal with removing carbon dioxide from the atmosphere on the ton-scale. Like with all trash that we humans produce, one method for dealing with carbon dioxide is to bury it in the ground. The problem with this approach is two-fold in that carbon dioxide would need to be condensed into a liquid to allow for this and that liquid would eventually change into a gas and find its way back to our atmosphere. A modified tack on this approach was reported in 2016, in the journal *Science*.[6] The researchers, located at a geothermal power plant in Iceland, took condensed carbon dioxide, incorporated it into sea water, and pumped the seawater through basalt rocks deep in the ground. They found that in two years time, all of that carbon dioxide was converted into calcium carbonate within the basalt. The researchers have scaled this process up, to attempt to sequester 5000 tons of carbon dioxide a year in basalt. These are very exciting findings as they are the first realistic method for the long-term removal of carbon dioxide from the atmosphere. Unfortunately, our human activities currently emit carbon dioxide on the scale of billions of tons every year. All 5000 tons that are being used in the pilot study, are being sequestered directly from the geothermal plant.

This gets to one of the problems that Alex and I talked about. He said, "The biggest problem with carbon dioxide conversion technologies, is that it's not easy to sequester. It's in the atmosphere, but compared to nitrogen, the concentration is really low." Even getting it from an industrial flue isn't straightforward. One of the most thoughtful things he and his collaborators are doing is developing catalysts for different natural carbon feedstocks. Catalysts for carbon dioxide where it's easy to get carbon dioxide (power plants). Catalysts for biomass where its cheap and easy to grow plants (seaweed and kelp in the Pacific Northwest). Alex's catalysts are being developed to make building block molecules from local feedstocks. Much like the local food movement (by Miller's description) you use the resources that are closest to you to create something of value.

We still have a long way to go with our work. But, I have great hope that we will be able to figure out a constructive way to reduce the excess amount of carbon dioxide in our atmosphere.

All of this research that I mention, all of it, is over the same reaction that is responsible for giving rise to many of the foods we eat.

It seems that there are several thematic reactions, several types of chemistries, that are unavoidable to humans: bicarbonate and carbon dioxide in the oceans and in our kitchens; photosynthesis and respiration and the consumption of fossil fuels for energy; forging iron just as the Earth's core was forged from iron. Do we rely on these reactions because they are the most efficient and effective? Or do we use these reactions because we are a product of the Earth, and the Earth is a product of this chemistry? I think that the answer lies more in the latter. We are born from this Earth and we employ the same devices that nature uses here for our own purposes, in our societies, in our industries, in our kitchens.

My pancakes (and your pancakes) are no exception to this. It is no surprise that we use sodium bicarbonate to leaven our pancakes with carbon dioxide. The sodium bicarbonate was once created by an ocean, churning its waters with carbon dioxide from the sky.

So, now I do my pancakes correctly. Any recipe that calls for baking soda requires the addition of something acidic (like buttermilk). Without it, both the flavor and the texture suffer. I actually prefer my pancakes a little bit thinner than in Madison's recipe. (Hers are very fluffy.) So I add a little more buttermilk to my recipe, making the batter a little looser. I compensate for the extra added acid by adding a little more baking soda and a little less baking powder. I don't have any precise measurements for this. It's just something that I've gained a feel for. But the differences in the soda and powder are only off by a minor amount in comparison to the recipe.

Another important part of Madison's recipe is the melted butter. In the past, I had always added a little pat of butter to a hot skillet before cooking my pancakes. It turns out that adding melted butter to the batter provides enough lubrication to help the pancake separate from the pan when you flip them. The added bonus to Madison's recipe is that your pancakes brown better (through the Maillard reaction) over the entirety of their surface. The browning that you get when putting your butter in the pan (instead of the batter) is not nearly as consistent or aesthetically pleasing.

Ever since I re-remembered my chemistry and started using buttermilk in my pancakes, we have made it a constant presence in our pantry. The acidity is important, of course. Lactic acid is the same acid that we feel in our muscles when we exercise, and is the same acid that we taste in sourdough bread. Just as sourdough bread benefits from the lactic acid produced by *Lactobacillus* bacteria, there are other flavors generated from fermenting dough with this bacterium. The same is true of buttermilk, which is fermented with *Lactococcus* and *Leuconostoc* bacteria.[7] Traditionally, buttermilk was fermented from the liquid that was left over from separating cream for butter making. The sourness of the acid is complemented by other flavors made during the fermentation.

Now we use buttermilk for all sorts of things around the house. Erika is the primary wielder of buttermilk outside of breakfast. She uses it to make ranch and blue cheese dressings. She uses it to make biscuits. And, in perhaps the ultimate calling for buttermilk, she uses it to brine chicken thighs for her fried chicken.

But, back to breakfasts since that's what I've really been talking about.

The other dish that I make for breakfast is Aretha Frankenstein's Waffles of Insane Greatness.[8] With a name like that, the waffles should be both good and pompous. They are. These waffles are everything that their name suggests. I had first heard about them from Molly Wizenberg's blog *Orangette*.[9] She was conducting a battle royale between different waffle recipes. The two finalists came down to Marion Cunningham's yeasted waffle recipe and Aretha Frankenstein's Waffles of Insane Greatness. Both of these recipes are really good. But when I want waffles, I make the insanely great ones.

In comparison to the pancake recipe, these waffles substitute cornstarch for some of the flour, decrease the amount of eggs, decrease the amount of leavening, and increase the amount of oil. [105 grams (3/4 cup of flour), 30 grams (1/4 cup) of cornstarch, 2 grams (1/2 teaspoon) of baking powder, 1 gram (1/4 teaspoon) of baking soda, salt to taste, 245 grams (1 cup) of buttermilk, 67 grams (1/3 cup) of vegetable oil, 1 egg, 7.5 grams (1 1/2 teaspoons) of sugar, vanilla to taste. Add each ingredient to a single bowl, mixing thoroughly after each addition.] The extra oil gets the edges extra crispy. The batter is thin. Leavening is not required as the little gas in the batter expands quickly in the waffle maker, leaving an airy interior.

A word of warning about topping these waffles: they are delicate. I love adding maple syrup to my waffles, but for these, I add only a scant amount. They will lose their crisp and get soggy after a little while. That is usually not a problem as they are never on the plate for too long. Two of our kids eat them plain. (They are that good. Seriously, what kid do you know who would pass up adding maple syrup to anything?) My favorite topping for these is homemade whipped cream. There is something about the cool cream contrasted with the airy crisp of these waffles.

On those mornings when Erika is treating me to a little down time, she makes one of her family favorites, David Eyre's pancakes, which are a variation of a Dutch baby pancake.[10] Erika's family calls them CCP's, after Craig Claiborne, a *New York Times* food critic who popularized the recipe after visiting Eyre in Hawaii. At the time, Claiborne wrote, "With Diamond Head in the distance, a brilliant, palm-ringed sea below and this delicately flavored pancake before us, we seemed to have achieved paradise."[10]

The recipe for this is a take on a popover recipe. Most of my introduction to traditional continental European cooking has come from my mother-in-law. She, whether on her own or through Erika, is the one who really introduced me to traditional French food, coupled with the joys of using ample wine and ample butter. For this pancake recipe, whose roots I also trace back to my mother-in-law, batter is made (no leaveners added) and cooked in a hot oven. The dough expands of its own accord and browns to a beautiful golden color along its crisp edge, leaving a more delicate custard in its center. [Preheat oven to 400 °F. After the oven is preheated, add a casserole dish containing 58 grams (1/2 of a stick) of butter until melted. Meanwhile, in a bowl, mix 2 eggs. Add 70 grams (1/2 of a cup) of flour and a pinch of salt, mixing loosely make sure that some clumps of flour remain. When the butter has melted, add the batter and bake for 20 minutes. Serve with powdered sugar and a lemon wedge.]

This recipe is also a big fan favorite in our house. The fresh lemon juice on top of the powdered sugar really makes it. And, even though they like squeezing lemon juice on their pancakes, my kids always show a little restraint, holding some un-squeezed lemon in reserve. Whenever we eat David Eyre pancakes, my kids, without fail, ask for baking soda. "Daddy, can we make carbon dioxide?" they ask. They love pouring lemon juice onto a pile of baking soda and watching the bubbles form and fizz. (We really can't escape this reaction. We even use it for entertainment!) It's a mess. But, I have to oblige. I'm doing a good job training my little chemists.

Perhaps the best part of making breakfast at home, in that time when I'm trying to give everyone a break, is when the kids come to help. They fight over who gets to measure the flour and who gets to add the sugar and who gets to crack the egg and break the yolk.

No, the best part of breakfast is being a dad and, without even trying, getting the kids to like the same things that you do.

REFERENCES

1. D. Madison, *Vegetarian Cooking For Everyone*, Ten Speed Press, Berkeley, 2007.
2. J. McLagan, *Bitter: A Taste of the World's Most Dangerous Flavor, with Recipes*, Ten Speed Press, Berkeley, 2014.
3. C. S. Tautermann, A. F. Voegele, T. Loerting, I. Kohl, A. Halbrucker, E. Mayer and K. R. Liedl, Towards the Experimental Decomposition Rate of Carbonic Acid (H_2CO_3) in Aqueous Solution, *Chem.–Eur. J.*, 2002, **8**(1), 67–73.
4. J. Abraham, *Climate Scientists Have Warned Us of Coral Bleaching for Years. It's Here*, The Guardian, 2016 June 10.
5. H. Arakawa, *et al.*, Catalysis Research of Relevance to Carbon Management: Progress, Challenges, and Opportunities, *Chem. Rev.*, 2001, **101**, 953–996.
6. J. M. Matter, *et al.*, Rapid Carbon Mineralization for Permanent Disposal of Anthropogenic Carbon Dioxide Emissions, *Science*, 2016 June 10, **352**(6291), 1312–1314.

7. H. McGee, *On Food and Cooking*, Scribner, New York City, 2004.
8. A. Frankenstein, *Waffles of Insane Greatness* [document on the internet]. Food Network; [cited 2016 July 5], available from: http://www.foodnetwork.com/recipes/waffle-of-insane-greatness-recipe.html.
9. M. Wizenberg, *You Deserve a Waffle* [document on the internet]. Orangette; 2010 May 18 [cited 2016 July 5], available from: http://orangette.net/2010/05/you-deserve-a-waffle/.
10. A. Hesser, *David Eyre's Pancake*, The New York Times, 2007 March 25.

Lunch

CHAPTER 5

Jelly

2.5 billion years ago.[1]

That is the earliest evidence we have for the onset of photosynthesis, the means by which living organisms use the energy from sunlight to convert carbon dioxide and water into oxygen and other chemicals.

This was a major turning point for life on earth. That these organisms were able to populate our atmosphere with oxygen is the most obvious change that led to our own existence. What may not be so obvious to many, however, is that these organisms were able to make their own food (energy source), becoming self-sustaining, not needing to scavenge from their environments for other energy sources.

One of the classes that I teach is general chemistry. In this class, which typically spans two terms, I try to drive home the importance of figuring out how much energy is stored inside of different molecules. This strategy is certainly not unique to me. Codified in nearly every introductory chemistry textbook are techniques and mathematical expressions for calculating energies. Certainly, there are some, strictly chemical reasons why we do this. Molecules with higher energies are more likely to react to form new molecules, if those new molecules have a collectively lower energy.

Chemistry in Your Kitchen
By Matthew Hartings
© Matthew Hartings, 2017
Published by the Royal Society of Chemistry, www.rsc.org

While this tenet is something that we chemical educators try very hard to help our students appreciate, it is a fact that life (and evolution) inherently values. Surviving, reproducing, and thriving, all require energy. Life extracts that energy from chemical reactions. In the example I noted above, where the higher energy molecules transform into lower energy molecules, the reaction also produces energy. This reality is an echo of the *First Law of Thermodynamics*: energy can neither be created nor destroyed.

And so, life finds ways to use energy. Most of the life on earth scavenges for its energy sources. There are microorganisms in the ocean that survive on the energy of hydrothermal vents. This environment is exceedingly inhospitable to most earthly life (high temperatures, toxic gases, toxic metals), but the bacteria that live near these vents evolved because of the compelling energy source. The ability to cultivate energy sources for living is on full display in humans. We hunt, we gather, we set up mass systems of agriculture, develop cold storage supply and shipping chains, and produce industrial fertilizer to facilitate food (energy) production.

The story of artificial fertilizer is a fascinating one (from a personal, historical, and chemical point of view). *The Alchemy of Air* is a remarkable book that chronicles this history, along with the heroic and tragic tales of the scientists who led the way.[2] In the late 1800's scientists made dire warnings about our ability to sustain the growing human population with the dwindling supplies of natural fertilizers that were needed for agriculture. Wars were being fought over islands and nations that had large deposits of bat guano! (It is telling that bat guano was used as both a fertilizer and a source for explosives – both necessitate molecules with high energies.) The scientific societies defined the grand challenge of this era as the conversion of nitrogen into ammonia, which could be used in the production of fertilizers. Nitrogen is cheap and abundant; it makes up nearly 80% of our atmosphere. But, it is a stable, low energy molecule. Converting nitrogen into ammonia was no easy task. And being able to perform this reaction cheaply and efficiently enough so that it could be scaled industrially, was even more difficult. The chemistry was first observed by Fritz Haber. The industrially efficient process was developed by Carl Bosch. Both men won Nobel Prizes for their work. Now, the Haber–Bosch process stands as, perhaps, the most important human-made chemistry on the planet. The ready access to fertilizer spawned by this chemistry facilitated the increase of our population by some 7 billion people. Some estimates claim that the reaction is directly responsible for one third of all human life.

As a species, we have gone, and will continue to go, through seemingly-extreme steps to ensure that we have the food (energy) we need to survive and be sustained. As remarkable as our own efforts are, they only put into context the ability of those species that are able to produce and store energy within their own cells.

Yes, photosynthesis does produce energy. But what happens if the organism doesn't need that energy at the same time as the completion of the photosynthetic reaction? And, what happens when energy is needed when there is no sunlight? These are questions that humans are facing with the development of higher efficiency solar energy panels: Where does energy come from when the sun goes down? The answer to these questions (in terms of both photosynthetic evolution, and in our current development of solar energy capabilities), is to store that energy in the form of chemical bonds.

Eventually, photosynthetic organisms evolved to produce simple carbohydrate sugars. Today, we formally define photosynthesis as the light facilitated reaction of 6 molecules of carbon dioxide and 6 molecules of water, to form 6 molecules of oxygen and 1 molecule of glucose. Most of us, no matter our scientific training, have a natural intuition for the energy content of glucose. As the parent of three children, I am well aware of how sugar can affect their activity (or hyper-activity). When we are feeling slow, there are two things that many of us reach for: coffee and candy.

Plants are no different to us in that glucose represents solar energy, stored within the bonds of a molecule. Within glucose, gifted by the sun, is the energy to live, to grow, to thrive, and to reproduce. Carl Sagan's quote, "We are made of star stuff," was meant to describe how every atom in our bodies was once made by a star.[3] The quote is accurate in an active sense as well. The sun's energy and photosynthesis continue to maintain our lives. This is true not only in the present, as we still benefit from the sunlight that hit the Earth millions of years ago. The oil and gas and coal that we rely on so much to power our lives, are the remnants, altered by time and geology, of ancient plants. Fossil fuels are the relics of chemical energy that we continue to harvest. The computer that I am typing on is fed by the sunlight of an Earth that knew no humans, translated into chemistry, and extracted by fire.

Almost every non-photosynthetic species on earth maintains a dependence upon photosynthesis. Our ability to extract energy from glucose, respiration, is so common to us that we often don't realize its connection to photosynthesis. Again, this is a fact that we are taught from a very early age. And, as an educator, it is a point that I try to hammer into my students. Respiration is the net chemical reaction by which we extract energy from glucose. The reaction goes as follows: 1 molecule of glucose and 6 molecules of oxygen, react to give 6 molecules of carbon dioxide, 6 molecules of water, and energy. Photosynthesis and respiration are mirror images of the same reaction. This chemical rearrangement is the blueprint that facilitates most of the life on earth.

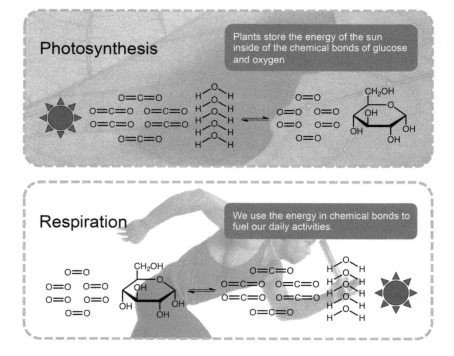

As nature started producing and using glucose, it produced other sugars (monosaccharides) as well. Glucose has the familiar molecular formula of $C_6H_{12}O_6$. There are other monosaccharides that have the same molecular formula. What separates glucose from some of these other simple sugars, is the order and shape in which all of these elements are strung together. By the demands of chemistry, photosynthesis doesn't faithfully produce only glucose. Chemistry is a promiscuous activity. Or, rather, chemicals are not shy about exploring different shapes and spaces during the course of reaction. While it turns out that glucose is the shape and structure most often made during photosynthesis, some other simple sugars are made as well.

After a while, nature got so good at producing simple sugars from sunlight that it figured out how to do other things with them, rather than just the short term storage of energy. There are three further steps in the evolution of carbohydrates, important for the food chemistry stories in this chapter, that I will discuss in more detail.

Glucose and the other monosaccharides are useful for short-term energy storage; photosynthesis produces a sugar that can be used later that day. Because of its easily accessed energy content, glucose will not last long within an organism. It will find some way to break down into carbon dioxide and water. Evolving plants had to find a way to store energy for times of need: during the winter or rainy seasons, when there is less sun, and during the early stages of going from seed to sapling, to name but a few. So, nature found a way to stitch monosaccharides together into long strands. Importantly, these strands needed to be easily and quickly snipped apart whenever the plant needed to access this energy. Long polymers of glucose molecules, which are easily accessed, has another name that we are almost certainly familiar with: starch. Cornstarch, wheat starch (flour), potato starch, tapioca starch, and many other kinds of starches, are all just the same thing: a way for these plants to store energy until needed, by connecting glucose molecules together into long strands. The differences between these starches come from how each individual plant packages them. Wheat starch, for instance, is packaged inside of a protein envelope. It is this same protein envelope that eventually turns into gluten when making dough.

Monosaccharides

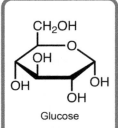
Glucose

Fructose

Ribose

Sulfoquinovose

Galactose

Galacturonic Acid

Methoxy-Galacturonic Acid

Plants turn sunlight into food by producing sugars (monosaccharides). The most important of these is glucose. Nature uses these molecules to store energy for long periods (amylose) and to provide protection and support (cellulose and others).

Mannose

Mannuronic Acid

Polysaccharides

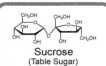
Sucrose
(Table Sugar)

A polysaccharide is just a monosaccharide that has been stitched together with a chemical bond. Because of their prevalence in Nature, they are also prevalent in our food. Many of those that we use for cooking (shown below and on the right) stretch for hundreds or thousands of monosaccharides.

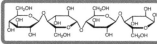
Cellulose
Differs from amylose in only geometry. Used for plant structure.

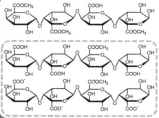
Pectin
Pectin makes jellies and jams set up. It is composed of mostly galacturonic acid (GA) and its variants. High methoxyl pectin (top), most commonly found at the grocers, has more methoxy GA. Low methoxyl pectin (bottom) has many negatively charged atoms.

Starch

Dietary starch comes in two chemical forms: amylose and amylopectin. Cornstarch contains a mixture of these two molecules. Starch is the most used polysaccharide used in cooking. Both are made purely from individual glucose molecules, strung together with chemical bonds.

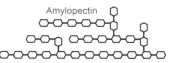
Amylopectin

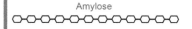
Amylose

Jelly

Photosynthetic organisms also used glucose as a way to make themselves stronger, to give themselves structure, and to protect them from their environment. The roots, stems (trunks), branches and leaves, all contain glucose-based polymers. The most common polymer on earth is a glucose-based polymer that you have probably heard of: cellulose. Cellulose allows trees to grow to such heights. Cellulose is also responsible for providing the plant protection all the way down to each individual cell. The glucose link in each cellulose strand must not be as accessible as the glucose in starch, lest the cellulose fall apart. Nature has found a chemical trick to ensure of this. A simple flip or twist on every other glucose within a cellulose polymer results in a material that is used for support, as opposed to energy storage. Starch is soluble in water while cellulose is not (thankfully, otherwise all of our cotton clothes would wash down the drain). Another noticeable difference between starch and cellulose is that plants tend to store starch in tiny bundles, folded away like origami, and cellulose remains extended.

The final trick that nature figured out, relevant to this story, is putting different chemical decorations all over those sugars. There are a number of reasons that this might happen. The environment that the plants live in may require different chemical functional groups. An interaction with minerals and metals, for example, would be assisted by the addition of groups called carboxylates or amines. The polymers may also have to adapt to support an environment that is either more, or less, hospitable to interacting with water. At any rate, small changes of the individual units of a polymer can have major aggregate effects on the polymer as a whole.

These observations, of all of the ways that glucose (and similar monosaccharides) gets used in biology, are representative of a larger trend: nature develops the tools to make something useful, and it finds ways to use those tools and those useful things, to become more complex. Glucose is the dominant molecule in biology because it performs so many different jobs.

LET'S JAM

For every way that nature has found to use glucose, there are countless recipes that humans have developed.

Certainly, we are partial to sugary foods. They taste sweet because our bodies recognize the need for sugar for energy. But, it is the more complex molecules that I am interested in here.

Cornstarch is a perfect example of this. Despite being an unpalatable ingredient (gritty and tasteless, it coats the tongue), cornstarch plays an assortment of roles in a variety of dishes. In popcorn, the cornstarch expands and fluffs up as it traps water vapor. The waffle recipe my family and I love (waffles of insane greatness from Aretha Frankensteins' restaurant in Chattanooga, Tennessee), uses a quarter cup of cornstarch for every three quarters of a cup of flour.[4] The waffles end up light and crispy, whereas the all flour version ends up chewy and dense.

Cornstarch is most often used to thicken sauces. As described earlier, cornstarch is a polymer, made almost entirely of glucose, bundled up into a tiny package. I like to think of it like a little balloon.

For Jack's 4th birthday, I filled over 400 balloons with water. All of those balloons were piled into a huge Rubbermade container. He and his friends and siblings had an epic water balloon fight in our back yard. It was glorious. It was also one of those times when I wish I weren't an adult. It'd be kind of inappropriate for a grown man to start forcefully throwing water balloons at little kids. Anyway, the balloons didn't fill immediately. They needed to be stretched and slowly filled. And, if they get overfilled, they burst and can't hold water any more.

Cornstarch is similar. All of the –OH groups interact with water; that little ball of cornstarch wants to be filled with water. But, it doesn't fill right away. It needs to be heated a little first. Once they're filled, they are a lot like the water balloons. A bucket of water with unfilled water balloons floating at the top sloshes around a lot like a bucket of water. But, if you fill the balloons with that water and tie them shut, the water in the balloons moves very differently. Collectively, it is more solid. Cornstarch is used to thicken gravies and puddings in much the same way.

A more complex polysaccharide (polymer of monosaccharides) that is no less useful, is pectin. Pectin is found in the cell membranes of fruits and vegetables. It breaks down into monosaccharides as the fruit ripens. Unripe fruit is firm because it contains a sufficient amount of pectin. Ripe fruit is soft because it no longer contains as much pectin. By the same token, there are some fruits that are naturally firm, which have high pectin content. These include: apples, crabapples, cranberries, currants, plums, quinces, and citrus rinds. There are some fruits that contain very little pectin. These include: apricots, blueberries, cherries, figs, peaches, pears, Italian plums, pomegranates, and strawberries. And, there are also fruits that have an intermediate content of pectin.

Instead of glucose, pectin is primarily composed of a monosaccharide called galacturonic acid. There are a couple differences between glucose and galacturonic acid. The biggest difference, chemically, is that galacturonic acid has an acid group ($-COOH$) attached to it. (Surprising, I know.) The other difference is the direction that the $-OH$ groups point (up or down) off of the ring structure. While galacturonic acid is a modified monosaccharide, it can be modified even further. Within pectin, many of these have a methoxy ester group ($-COCH_3$), instead of the acid group. The difference between acid and ester makes a really big difference when you are using pectin in your cooking.

And, the way the pectin is most often used is in making jams and jellies. A quick set of definitions are in order here. Both jams and jellies are made from cooking fruit with pectin. Jellies are strained before they set and end up firmer, clearer, and smoother. Jams are not strained and are looser, less smooth, and often contain visible chunks of fruit.

Perhaps it is best, at this point, to talk about why foods like pudding and jam and jelly all have different textures. They are thickened by polysaccharides. But, it is the small differences between the polysaccharides and the variations in food preparation, that drive the nuance between the emergent textures while cooking.

The textures that I am describing are solids and liquids and all of the various degrees between these two extremes. The texture is a result of how fast individual molecules within a material move and their range of motion. One of the things that I love most about chemistry is that you can tell what's going on with something as small as a molecule, just by watching a huge collection of molecules. Take water, for instance. Liquid water is nearly free to move around wherever it likes. Pour water into a cup and the individual molecules are able to zip around, changing direction only when they run into another molecule, or into the side of the cup. They can't get out of that cup, however. They are also bound by the surface of the liquid; the water molecules are too attracted to one another to be able to leave these confines. External forces can change the natural ebb and flow of that liquid water. We can pick up that cup and pour the water into our mouths. Sometimes we miss our faces and the water ends up all over our clothing. And, as we've seen in *Jurassic Park*, a passing *Tyrannosaurus Rex* will cause the water in a cup to reverberate in perfect, concentric circles. (Note: this bit of movie magic was actually caused by plucking a guitar string underneath the cup of water.)[5]

Give that water more energy, however, and it is no longer bound by an open container, or itself. When we boil a pot water, the heat energy from the stove is imparted to each water molecule inside of that pot. Once the water gains enough energy, it is able to break free from these constraints and go hurtling off to explore new surroundings. Much like how our own rocket ships need to reach escape velocity before they can overpower the pull of gravity, water molecules need to move fast enough to change from a liquid into a gas.

We can understand water molecules in ice then, as water molecules that have slowed down. Remove energy from liquid water, and its molecules stop moving. In the ice that we know from our daily lives, motion has effectively stopped. There is still some wibbling and wobbling going on, but the molecules no longer move up or down or left or right. They are stuck in place, without the energy to break free from the attraction to their neighboring molecules.

Texture, then, can be thought of the ability of the molecules within a material to move. Ice is firm and hard precisely because its molecules no longer move around. Liquid water is, well, a liquid precisely because its molecules are free to move. Foods that do not flow as well as water but are not as firm as ice have molecules whose motion falls along an intermediate spectrum. The more motion in its molecules, the more fluid it will be. The slower motion (or more restrained motion) of its molecules, the more firm it will be.

Taking this lens and focusing it on pudding and jam and jelly, we can start to understand how the polysaccharides are operating. (Another jargon-y term that describes a polysaccharide is "hydrocolloid". This is the term that is most used by the food community and so I will mostly use hydrocolloid for the remainder of this chapter. A colloid is a large particle (well, much, much larger than a flavor molecule or a sodium ion, anyway). Colloids can be suspended and dispersed throughout a liquid. A hydrocolloid is a particle that has strong interactions with water. As the polysaccharides that we are talking about here are large particles that interact strongly with water, they can be accurately described as hydrocolloids, of which there are many varieties.)

Jam (uses pectin) and pudding (uses cornstarch) have a similar texture and are firmer than water but less firm than jelly (also uses pectin). In jam and pudding, the hydrocolloids absorb water and swell like balloons. The water, interacting with the hydrocolloids, slows its motion. Consider a situation where you are walking down the street, moving at your own pace, and, all of a sudden, the entire sidewalk is covered by discarded pieces of bubble gum. You are still going to continue walking, just at a slower pace as the gum grabs hold of the bottoms of your shoes.

What, then, is the difference between the way pectin works in jelly *versus* the way it works in jam? Why is jelly noticeably firmer than jam? The answer to these questions, it turns out, is in the way the hydrocolloids interact with each other. The interaction between water and the hydrocolloid is enough to slow down the water. But, when the individual hydrocolloid polymers start interacting strongly with one another, then the mixture can set into something more firm.

The reason why Jell-o, which uses gelatin as its hydrocolloid, has the consistency that it does, is that the gelatin molecules interact strongly with one another. Individual gelatin particles grab on to each other and make a web, or network, that runs through the entire volume of the mixture. They attract water, trap it, and also trap themselves. The transition from squishy to firm, then, is all about controlling the motion of the hydrocolloid.

In jam there is too much fruit floating around to allow this interaction to happen. The pectin molecules will reach out, and most likely interact, with a piece of fruit. Contrast that to jelly, in which the liquid gets strained through a filter. In this scenario, one pectin particle can stretch its arms and find another pectin to grab onto.

Making jellies or jams for some fruits is very straightforward. There are two requisite things that the fruit has to have: pectin and acid. It is fairly obvious that a fruit with a low pectin content will not set into a jelly. The acid necessity, which we will discuss later, has to do with the way that one pectin particle grabs hold of another. Cranberry jelly is pretty easy to make. Add cranberries a little water and a little sugar (and maybe some lemon zest), to a small pot. Heat the cranberries until they have popped open and softened. Strain and cool.

Things get a little more complicated for fruits that don't contain a lot of pectin. Unfortunately, these include some of the most popular jams and jellies (strawberry and raspberry) which do need added pectin. So, what is a jelly-craving person to do. Well, for the adventurous home cook, you can make your own! You can extract pectin from unripe apples by boiling those apples in a bit of water.[6] You can also very easily purchase pectin from the grocery store. But, you've got to be a little careful when you do because there are several kinds, and their chemistries are just different enough to be important.

Now, if you are the type of person who just follows the directions listed on the back of the pectin packet, you'll be fine. However, understanding the chemical differences between these two can open up a lot of possibilities for your cooking.

The two kinds of pectin that you can buy are high-methoxyl pectin and low-methoxyl pectin. High-methoxyl pectin is usually just labeled "pectin", with no other qualifiers. Low-methoxyl pectin is usually labeled "low-sugar/no-sugar pectin" or "universal pectin".

The difference between these two kinds of pectin is a small one. Remember the galacturonic acid that makes up the pectin polymer. For many of these, the links in the chain, galacturonic acid, is actually the methoxy ester variety and not the acid variety. For high-methoxyl pectin, an overwhelming number of the links are not acidic. For the low-methoxyl pectin, more of the chain links are acidic. In other words, low-methoxyl pectin has more acidic groups than high-methoxyl pectin.

And that acidity, coupled to the charges that go along with it, plays a big role in how we use the different kinds of pectin. In the chapter on cheese, we will discuss how acids like to lose protons (H^+) and become negatively charged. That means that high-methoxyl pectin and low-methoxyl pectin will both be negatively charged. This is a major problem if we want the pectin particles

Jelly

in our jams and jellies to link up. Negative charges repel one another. And the pectin particles, having many negative charges, will never interact with one another, and will never set, as long as they are negatively charged.

Fortunately, there is another trick in the cheese chapter that can help us with this quandary. All acids are a little rebellious. If there are other kinds of acid floating around them and giving up their protons, the acids on pectin will decide that they'd rather just keep their own protons. Put enough of an extra acid in with the liquid that you are trying to turn into jelly, and the pectin will cease being negatively charged and will start interacting with one another, and set.

This step is in the direction of making jams with high-methoxy (regular) pectin. The addition of a little bit of lemon juice to the mix, on top of the other acids that are naturally present in the fruit that you are using, is enough to keep the pectin polymers from being negatively charged. Along with the lemon juice and fruit, the other key ingredient in a typical jam recipe is sugar. Lots and lots and lots of sugar.

The sugar plays several key roles in these recipes. It is there for flavor. (Everybody likes sugar, right?) An excessive amount of sugar helps to slow the growth of bacteria and mold and helps to keeps your jam from going bad. The other important thing that sugar does is that it assists the pectin in getting the jam to become firm. High-methoxy pectin, on its own, doesn't have the ability to slow down water molecules to get the consistency expected for a jam. The sugar helps with this. We see this in action with syrups, which become thicker and oozier the sweeter they become.

But, what happens if you don't want a super-sweet jam or jelly? What if you want a jelly that tastes more like the fruit you're trying to use rather than a flavored sugar bomb?

This is where the low-methoxyl pectin comes into play. You can make jams with low-methoxyl pectin using the same recipe as with high-methoxyl pectin. But, you don't have to. And the reason why you don't have to is because of all of those extra negative charges. Those charges that would normally make the pectin particles repel one another can be used to your advantage.

While negative charges will repel one another, a positive charge and a negative charge will attract. If you throw a big enough positive charge into the middle of it all, it can attract several negative charges to it. That's basically how you get low-methoxyl pectin to set into a gel. Cook some fruit until it starts to break down. Add some low-methoxyl pectin. Add some calcium ions. Calcium has a +2 charge. It can pull in a negative charge from one pectin polymer and another from a second, basically facilitating the thickening of the gel and, voila, giving you a jam that tastes just of the natural fruit.

The chemistry that you use for one type of pectin isn't necessarily better than the chemistry that you use for the second (unless you're on a low sugar diet). The chemistry gives you options. Right there in the grocery store, you can find prepackaged, premixed hydrocolloids that will firm up into different kinds of gels that fit the types of food that you want to make.

MOLECULAR SCHMOLECULAR

Molecular gastronomy burst onto the culinary scene when Ferran Adrià began serving fruit caviar at his restaurant elBulli. This dish, served in a caviar tin, was such a hit because its taste completely contradicted the expectations of the diners eating it. They expected ocean brine. They got sweet and fruity.

The key to this dish is a hydrocolloid called sodium alginate. It is a polysaccharide that comes from algae, as you could probably guess from the name.

Adrià and other chefs bristled at the name "Molecular Gastronomy". Many prefer "Modernist Cuisine" instead. The philosophy around this movement involves cooks using the best techniques available for innovative food preparation. While my description seems bland and borders on meaningless (seriously, who wouldn't use the best techniques), the chefs at the forefront of these trends really pushed to find new techniques, new foodstuffs, and new points of view. In other words: the best means possible, from any means possible. The eponymous, encompassing, and beautiful book, *Modernist Cuisine*, helmed by Nathan Myhrvold, takes this view to heart.[7] Myrhvold describes and illustrates the science behind a myriad of cooking methods, from the mundane (baking) to more esoteric processes (using *trans*-glutaminase to glue pieces of meat together).

The phrase "Molecular Gastronomy" was coined by physicist Nicholas Kurti and chemist Hervé This. Many chefs aren't fond of this description because they think that it sounds elitist and inaccessible. I don't think that molecular really encompasses the full breadth of techniques spawned by this movement. But I personally think that the name stuck because many of the memorable advances and expressions were brought about by the use of new food molecules that facilitated mind and palate-bending recipes. (This name is also bolstered by the use of equipment traditionally found in a chemistry lab: centrifuges and rotary evaporators, to name but a few.)

Foams, fruit caviars, raviolis, spherified olives, powders, and gels, are just some of the dishes that caused a sensation.

All of these foods are only possible because of novel uses of hydrocolloids.

Much like how jellies and jams require precise amounts of fruit, pectin, sugar, and, sometimes, calcium, these modernist dishes also require precise amounts of hydrocolloids and other ingredients. What's different is that the pectin (and sometimes calcium) are pre-measured for you. When you make jam, you just dump in the packet. If you want to make fruit caviar, you need to have the mass of the juice, it's pH, the mass of alginate, the mass of water, and the mass of calcium, all worked out.

It can be intimidating.

In reality, this isn't true at all. Making some of these dishes is much more straightforward, and less difficult, than many other things that we make in our kitchens. In fact, much of the chemistry behind the preparations in molecular gastronomy is, to me, a little dull. One of the draws of cooking chemistry is the complexity of everything. It's dirty. It's raw. I find beauty in that. Molecular gastronomy, on the other hand, is very clean and prescribed and sterile. If you have a good kitchen scale, you can make most of these recipes.

The biggest problem that I have found in reproducing some of these molecular recipes, is getting the flavor intensity right. Because of their reliance on hydrocolloids, the flavor can be a bit muted. Many hydrocolloids will coat your tongue and cover your taste buds. When they get used in larger quantities, as they are in these recipes, they are more likely to dampen flavors.

It can be difficult.

Luckily for us, we have Martin Lersch to show us the way. Martin Lersch is a PhD organometallic chemist from Norway. Much like me, he studies metals and the reactions that they facilitate. Also, like me, he is fascinated with food chemistry and fascinated with the newest cooking techniques. Unlike me, Martin has made real contributions to helping people make modernist cuisine recipes at home.

In his spare time, he runs a blog called *Khymos*,[8] which is Greek for juice, but which also obviously shares a root with chemistry. Martin's blog has mostly been a collection of musings about modernist cuisine and exploring how to make different recipes reproducible.[9] That is the scientist in him coming through. A one-time success in the kitchen (or in the lab) does not mean you have mastered (or understand) the system. All great chefs and scientists strive for reproducible greatness or reproducible novelty. The entirety of *Cook's Illustrated* is devoted to finding the recipe variations that work the best every time. The problem with repetition of novelty is that it renders novelty ho-hum.

That is a tension that lies at the heart of high-end science and high-end cuisine. You build up a field or a technique, and find that the edge has moved. It is the opposite of erosion. Staring over the precipice for a long time, watching carefully, probing its boundaries, you find new edges to move to that are farther away from what you once thought was the end.

With a scientist's eye for detail and a keen understanding of food chemicals, Martin set about writing a guide to help the home cook jump into the world of molecular gastronomy. He has put together a brilliant guide, *Texture*, that I recommend to anyone whom I think might be interested in this style of cooking.[9] His guide introduces a number of hydrocolloids and other modernist food additives (agar, carrageenan, cornstarch, gelatin, gellan, guar gum, gum arabic, isomalt, konjac, lecithin, locust bean gum, maltodextrin, methyl cellulose, pectin, sodium alginate, and xanthan). He describes their uses and tolerances (what other ingredients they can and can't be used with) and the limits of what they can do. He also describes how to prepare these hydrocolloids for use in food. As many of these chemicals come in powder form, they need to be hydrated before use. Each needs its own technique for hydration, and it can be confusing to decide when to use one over the other. Lersch has taken care of all of that for us. But, he goes a step farther, and his extra steps are what make this an indispensible guide.

Cornstarch

Hydration and Gellation

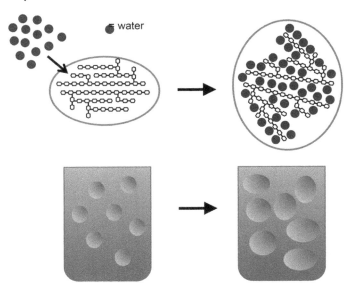

Weak Gel   Strong Gel

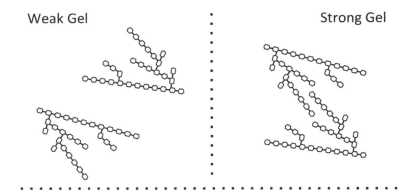

Texture contains a number of recipes that have been crowd-sourced and peer reviewed. The readers of *Khymos*, as interested and adventurous as Lersch, submitted their own recipes that highlight the use of each hydrocolloid. The peer reviewed part deals with quality control. Lersch made sure that he, or someone else, tested the recipes that were submitted for his collection to make sure that they work. (This is a step beyond peer review in science, where we only read and analyze the study. Sadly, we don't have the time or resources to try all of the experiments that are detailed in a research paper.) These recipes will work in your home kitchen; they are not some out-there ideas from an *avant-garde* chef. If you are curious, check them out.

The other source that has influenced how I think about hydrocolloids in food is the book *Ideas in Food* by Aki Kamozawa and H. Alexander Talbot.[10] These chef-authors had worked in professional kitchens when they started their own culinary consulting business. Much of their efforts were aimed at helping chefs learn and use new techniques in their own restaurants. As such, they have a great deal of expertise in the modernist approach. (There is something about teaching. It actually forces you to learn the material in much greater depth than when you only need to apply that knowledge in practice.) At any rate, this understanding really comes out in all of their books. *Ideas in Food* does a fantastic job describing hydrocolloids in food and the textures they can bring. They actually taught me a bit of chemistry in this book. I highly suggest it if you are looking to start getting adventurous with your cooking. They've got a lot of really playful recipes that will make you catch the bug for expanding what you do in the kitchen.

JUST JAMMING

In some ways, our culinary expressions are just another step in the evolution that turned sunshine into sugar. Just as biology figured out all sorts of uses for sugars, as ways to store energy, and as building blocks for making bigger, structural molecules, we are finding new ways to exploit their chemistry into making tasty and creative foods. I don't think that we'll ever run out of innovative ways of using nature's bounty, and I know that, if we do, nature will always surprise us with something else that's new.

REFERENCES

1. J. M. Olson, Photosynthesis in the Archean Era, *Photosynth. Res.*, 2006, **88**(2), 109–117.
2. T. Hager, *The Alchemy of Air: A Jewish Genius, a Doomed Tycoon, and the Scientific Discovery that Fed the World but Fueled the Rise of Hitler*, Broadway Books, New York City, 2009.
3. C. Sagan and J. Agel, *The Cosmic Connection: An Extraterrestrial Perspective by Carl Sagan*, Anchor Press, Garden City, New York, 1973.
4. A. Frankenstein, *Waffles of Insane Greatness* [document on the internet]. Food Network; [cited 2016 July 5], available from: http://www.foodnetwork.com/recipes/waffle-of-insane-greatness-recipe.html.
5. I. Phillips, *An Earth, Wind, and Fire Song Inspired Steven Spielberg to Create one of the Most Terrifying Scenes from 'Jurassic Park'*, Business Insider, 2015 June 5.
6. Benivia, *Make Your Own Natural Pectin for Use in Making Homemade Jam and Jelly* [document on the internet]. Pickyourown; [cited 2016 July 5], available from: http://www.pickyourown.org/makeyourownpectin.htm.
7. N. Myhrvold, C. Young and M. Bilet, *Modernist Cuisine*, The Cooking Lab, 2011.
8. M. Lersch, *Khymos* [document on the internet]. Khymos; [cited 2016 July 5], available from: http://blog.khymos.org.
9. M. Lersch, *Texture* [document on the internet]. Khymos; 2014 February [cited 2016 July 5], http://blog.khymos.org/recipe-collection/.
10. A. Kamozawa and H. A. Talbot, *Ideas in Food: Great Recipes and Why They Work*, Clarkson Potter, New York City, 2012.

CHAPTER 6

Macaroni and Cheese

I loved Kraft Macaroni & Cheese when I was a kid. Actually, I still love it, but it might be due more to nostalgia than anything else. There are two distinct times when my brother and sister and I would be served Kraft Macaroni & Cheese: during lunch on weekends, and in the summer. What kid doesn't love time outside of school? I couldn't help but be endeared to the stuff. I still remember when I finally started making it myself (sometime in my early teens). I love it when the "sauce" is a little runny, so I would add extra milk to it. Heavenly. I also remember the first time I ate Kraft Macaroni & Cheese with Erika. She seasoned hers with pepper and ate it with a fork!! (She was so much more refined than I was ... she still is!)

Kraft made a big splash recently by removing the artificial ingredients, specifically some of the dyes. In the States, we are used to our Mac&Cheese being a vibrant, glowing yellow-orange color. The dyes, yellow 5 and yellow 6, were replaced with paprika, annatto, and turmeric. Kraft could have just made a decision to not color their product, but we Americans are pretty darned attached to our nostalgia. What didn't change is the part of the powder packet responsible for making the sauce sauce-y: whey, starch, whey protein concentrate, cheddar cheese, and granular cheese. (Whey is the liquid part of milk that is leftover after making cheese. It consists of water and several proteins.)

Any macaroni and cheese sauce needs to: taste like cheese, coat the pasta, and have a mouthfeel that is smooth, creamy, and thick. This is easier said than done. There is a reason why people use Velveeta. My European readers are probably snickering at us unsophisticated Americans right now. There is a broader appreciation and expectation for good cheese in Europe than here in the States. However, I'm sure that many American readers are snickering at me right now too. But, let me say this for Velveeta: It melts, it coats, and it tastes really good with cheap tortilla chips and corn chips.

Velveeta does contain some milk and cheese-style ingredients: milk, whey milk protein concentrate, milkfat, whey protein concentrate, and cheese culture. By deconstructing milk and putting it back together you can control the flavors and melting profile of cheese while increasing its shelf life. A similar process happens with chocolate. Cacao nibs are difficult to source because they don't stay fresh for very long. Cocoa powder and cocoa butter, on the other hand, are stable for many years after they are separated from the nibs. Furthermore, the recombination of cocoa powder and butter to make chocolate, results in a product that can be stored for a considerable amount of time.

I'd like to add one other note on Velveeta. There is a whole sector of modernist cuisine, the movement that puts high tech gadgetry and ingredients to use in making new foodstuffs, that is devoted to deconstructing and reconstructing foods. In an alternate universe where Velveeta had never been invented, the modernist chefs would certainly be trying to make their own deconstructed/reconstructed cheese. If one of these chefs were to make something like Velveeta, they would be lauded by food critics and high-end diners alike. The science and effort that went into making Velveeta is no less extraordinary because it was developed in a food industry laboratory.

The benefits of long shelf-life and meltiness come at a cost. We are not crude matter with no ability for discernment. Velveeta decomposes cheese into necessary components. It is a blunt instrument with no nuance. Real cheese, good cheese, livens our senses because it is at once assaulting (strong flavored) and remains tinged with a quieter, complementary fullness.

Making a cheese sauce with that rounded out flavor requires a cheese that melts in just the right way. There is a reason why Gruyère gets used so often in making cheese sauces, which includes both fondues, and macaroni and cheese. But, when I want mac and cheese, that usually goes along with a craving for cheddar.

If you have ever melted cheddar before, you'll notice that it turns into an oily mess. Tiny clumps of solid are seen floating in a greasy, oily liquid. And, it tastes like it looks too: gritty and disjointed. All of this happens because of the way that the proteins in milk come together to form cheese.

MILKY MUSINGS

My kids are on the swim team at our local pool. The end of May always brings excitement as the kids are looking forward to the end of school and the start of swim team, where they get to hang out with their friends and just be kids for a bit. This enthusiasm is usually dashed when we have one of those unseasonably cold patches of weather. (It's funny to call it unseasonal when it happens every year.) The trips to the pool, then, correspond to swimming in cold water, and kids complaining. A lot of hot chocolate is made to help us get through it all.

I'm kind of finicky about the way I make my hot chocolate. I'm not finicky about the ingredients or the quality of the chocolate or anything like that. I'm finicky about absolutely not having any of that nasty film that sometimes forms at the top of milk while it cooks. I really despise that stuff. It's not so difficult to prevent that film from forming; it just takes a little stirring as you heat up the ingredients.

But the milk film really is strange, isn't it? The why and the how of that film forming has to do with the fact that milk, and the proteins it contains, are special. For being a food with a lot of protein, milk doesn't act like other foods that contain a lot

Macaroni and Cheese

of protein. Specifically, milk proteins don't respond to heat in the same way that most proteins respond to heat. Although we've talked about eggs in an earlier chapter, let's talk about them again to illustrate what happens as we cook the proteins in our food. Initially, the egg takes on a liquid form. In this liquid, there are all sorts of proteins and minerals and other things swimming around in a big chemical party. The proteins are shy and introverted, wrapping themselves up into little crumpled masses, trying to hide amongst all of the water. As the party heats up, the proteins start to stretch out and interact with the other chemicals. Eventually, all of the proteins open up, grab onto each other, turning the egg from a liquid to a solid.

Milk proteins do not operate in the same way. Heat them up, and they continue to keep to themselves. Except for at the surface. At the surface of your milk, when water quickly evaporates as you heat it, the proteins find themselves without any water to protect them. They are forced to interact as the water disappears. When heated milk is stirred enough, the water from the middle and the bottom of the container replenishes the water that is evaporating from the surface. And, any proteins that had been hanging out at the surface, find themselves continuously surrounded by water, not forced to be social.

What makes milk different can also make it useful. Evaporated milk and sweetened condensed milk are the products of milk's heat tolerance. Both products are useful in the kitchen. In the past, evaporated milk was especially important for providing a stable source of nutrients that don't spoil.

If the proteins in milk don't cook the same way as the proteins in egg, and if cheese is just a hardened form of milk, how does cheese get made?

Milk proteins come in two different varieties. There are the casein proteins and the whey proteins.

Whey proteins are a set of proteins that are soluble in water. They are happy enough just hanging around and doing their own thing. They ball themselves up into tight little structures, like other good proteins. Nothing too special about any of these proteins on their own.

Casein proteins are different. An individual casein protein doesn't exhibit a whole lot of structure. It stays rather extended in comparison to other proteins. For any protein, being extended is a

Chapter 6

Milk

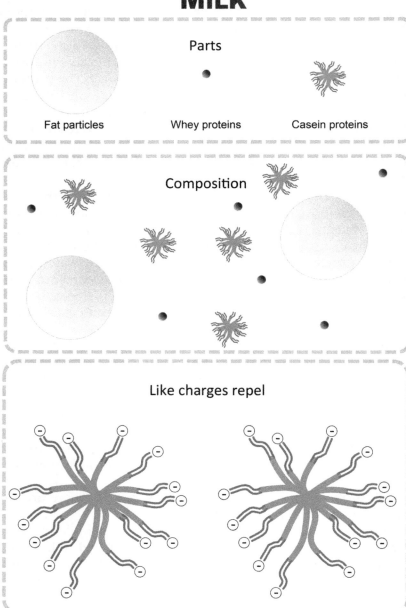

quality that does not bode well for its survival. Extended proteins don't tend to be soluble in water. Extended proteins don't tend to be useful to biological organisms. In fact, biological organisms have evolved another set of proteins to deal with, and get rid of, extended proteins. Our cells are filled with proteins whose sole job is to destroy other proteins, proteins like casein.[1]

But casein is tricky. It gets around all of this by exploiting what it is. And what it is, is a stretched out protein that's got one end that would rather be in oil and one end that would rather be in water. So casein does what any chemical that looks like a worm with two ends that would rather not be around the other does. It finds more worms just like it, and they line up with their water-loving ends hanging out and their oil-loving ends hanging out. When these proteins are in water, which is the milieu that we all work in here on Earth, the oily ends point in and the watery ends point out. The end result is something that looks like a Koosh ball.

In the grand scheme of things a Koosh ball is a structure. And biology loves structure. The biological degradation machinery won't touch it. If it were just a worm, it'd be chopped up into wormy little pieces. But as a Koosh ball, with many proteins working together to avoid a worse fate, casein survives.

All surviving proteins are usually given a job by nature. Or, more correctly, nature finds all proteins that have a job, a way to survive. The casein Koosh ball's job is the responsibility of all of those little strings sticking out of the center of the ball. It turns out those strings are good at carrying calcium ions around.

The easiest way to latch onto calcium ions is with negative charges. In water-based systems, like milk, calcium has a charge of +2. We scientists assign things arbitrarily, or, at least it seems that way some times. But we try to be consistent with our arbitrariness. The electric charge of any chemical is based on the charge of an electron. We have arbitrarily assigned an electron a charge of −1. When calcium goes from its elemental state (no charge) to its ionic state (+2), it literally loses two electrons. So, if you start off at 0 (no charge) and you lose two electrons (−2), you are left at +2.

At any rate, calcium has a +2 charge and is attracted to negative charges, which casein has in spades. In many proteins, negative charges come from the amino acids aspartate and glutamate. But, in casein, the negative charges primarily come from phosphate groups. These particular phosphate groups were put on casein by nature just so the protein could swallow up a bunch of calcium ions.

So, we need to make one more modification to our Koosh ball-as-casein analogy. The ends of each strand, sticking out into water, contain lots of negatively charged phosphate groups. As those negative charges all repel one another, the Koosh ball ends stick straight out. The effect is the same as when you rub the top of your head with a balloon, getting it all static-y. The effect is the same because the physics of each are the same. The balloon-assault puts excess electrical charges all over the strands of hair. The charges, and the hair, repel one another, and you end up looking like you stuck your finger in an electrical socket.

With each casein strand all charged up, the casein-Koosh balls collectively have loads and loads of charges. It follows that each of these Koosh balls, floating around in their milky solution, also repel one another.

In cheese making, the milk gets transformed into two substances. Solid clumps that get formed into shapes and eventually become the cheese that we eat. The other component that remains in the cheese making process is a runny liquid called whey. Many of you will remember Little Miss Muffet who sat on her tuffet, eating her curds and whey. The whey contains water and all of those whey proteins we discussed earlier. The curds contain the milkfat and casein proteins. Cheese making, it seems, negates the effects of all of those negative casein charges.

The first step in cheese making involves warming the milk. But, as we've seen, that has very little effect on the milk proteins. A second step is critical. Acid needs to be added to cause the formation of curds.

Making cheese curds is really easy to try at home. It also happens to be one of my favorite demonstrations to do in class. The recipe I like to use is from Ricki Carroll, founder of the New England Cheese Making Supply Company.[2] Start off by making a solution of citric acid in water, 2 teaspoons in 1 cup of water. (If you don't have citric acid at home, you can also use vinegar or lemon juice. Citric acid gets used because it lends a cleaner flavor to the cheese. Translated: you can taste the vinegar and the lemon in the cheese curds after you make them.) Slowly heat a gallon of milk with half of the citric acid solution. After a few minutes, you should see flakes (the start of curd making) floating to the top of the milk. If this isn't the case, add a little more of the citric acid mixture. Slowly stir until the milk reaches 88 °C (190 °F), and then move the pot to a cool part of the oven. Gently move any curds to the center

Macaroni and Cheese 107

of the pot and allow them to slowly meld together over 10 or 15 minutes. Remove the curds and let them drain.

What is it about acid that causes this? Why does acid "cook" milk proteins when heat won't?

There are other dishes that use acid, and not heat, for preparing proteins. Of these, ceviche is probably the most well known. Ceviche is prepared by mixing fresh seafood with lemon or lime juice, and letting the mixture sit for several hours. As the fish is exposed to the acid in the citrus juice, the texture noticeably changes. The acid induces unfolding and aggregation similar, but slightly different, to the way that heat does when cooking.

You can cook an egg with acid too. Let an egg sit in vinegar for a day and it will become firm and rubbery.[3] (This is usually a crowd favorite. My oldest daughter came home from school recently raving about doing this science demonstration at school.)

But cheese works a little differently than ceviche. The proteins in casein are already unfolded and have already aggregated. The question is: how do you get a bunch of negatively charged particles to come together and solidify.

ON AND OFF

I'm going bald. This is no surprise to the people who know me. It's been a long time coming. It started to become noticeable in my senior year in high school when I was 17. I was a football player. American football. My hair loss probably started from pulling my helmet on and off several times a day. (I don't blame football for my lack of hair. It was a foregone conclusion that I was going to lose it at some point.) Anyway, in the summer before my senior year, I also got a little abrasion at the leading edge of my hairline from my helmet. This abrasion ended up turning into a scar. So now I have a constant reminder of where my hair used to be.

I am sure that you're asking what this all has to do with cheese. Trust me, I'm getting there.

I tend to be a little self conscious about my scalp on sunny days. Part of it is that I'm worried about overexposure and sun damage. The other part is that I'm worried about other people getting hit by the glare from my pate. There's already enough sunshine around. Folks passing by don't need the extra light bouncing off of my dome. So, on sunny days, I wear a hat. The question I have to ask myself in the morning is: how sunny is too sunny? Well, now that I've been losing my hair for over half of my life, I've gotten pretty good at knowing where that line is.

Getting back to cheese. Let's take a closer look at those phosphate groups in milk. They're around because they can be negatively charged. But, they don't have to be. In fact, there are situations when they aren't. A phosphate ion is a phosphorus atom surrounded by 4 oxygen atoms. When a protein is phosphorylated, one of those oxygen atoms becomes bonded to the protein. Two of the three remaining oxygen atoms can have a negative charge. What dictates the presence of the negative charges is how much acid is present in the water.

When an acid is added to water, it gives off a proton, which we chemists symbolize as an H^+. Once there are enough protons swimming around in solution, the phosphate can't help but grab a hold of some. When that happens, the protons head directly to the negatively charged oxygen atoms on the phosphate.

Just like how I need to put on a hat when its sunny, the phosphates put on a proton when there are so many protons already in the water. And, just like I use my judgment to figure out how

PROTONS & CHARGES
& PROTEINS

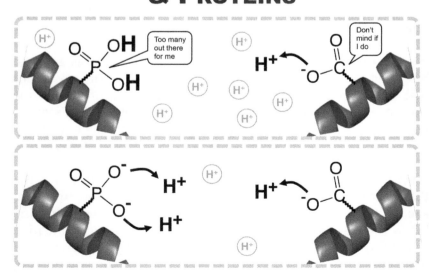

sunny is too sunny, the phosphates can tell how many protons are too many.

So the phosphate groups on casein take on some protons. They go from being negatively charged to having no charge. Now, instead of repelling one another, the Koosh balls can stick together.

And that's what cheese curds are. They are casein proteins, which have lost their negative charges and have decided to stick together. When they do this, they trap some of the milkfat. The liquid that gets left behind is the whey. Little Miss Muffet's tuffet-top dinner was some curdled milk and the watery liquid that is the result of adding acid to milk.

In Ricki Carroll's cheese curd recipe that I described earlier, this is exactly what is going on. You add enough extra acid to the milk to get the amount of protons past casein's make or break point. (In scientific terms, the amount of protons in solution is reflected by the pH; the higher the concentration, the lower the pH. The make or break point is called the pK_a. The pK_a is just the specific pH at which casein starts putting on protons. Every chemical has its own distinct pK_a.) When I make this recipe at home, I use the curds as a stand-in for ricotta. It's not real ricotta, as that is traditionally made with the whey. But, it's close enough. It is easy to make and tastes delicious. We use it when we make lasagna. It is good on its own as well. We like it warm on water crackers with a little bit of honey. Delicious!

While these curds are tasty, they don't really have the requisite internal structure that is required for making most cheeses. The consistency of this fresh cheese does not easily elicit comparisons to hard cheeses, like Parmesan. And there is a reason for that. The interaction between casein particles is ill defined and somewhat haphazard. The particles are held together by forces that are less than permanent. In chemistry, most of us tend to think of covalent attachments as being permanent. They aren't, really, but they are certainly more permanent than the fleeting interactions that bring curds together.

To make cheeses that can stand up to the process of aging, the casein interactions need to be stronger. We have established the Koosh ball shape of casein. For our way of thinking, a single colored Koosh ball implies that there is only one protein that makes up casein. In actuality, we should be thinking of a multicolored Koosh ball. There are several proteins and several smaller molecules that are part of casein. One of the most important proteins is called *kappa*-casein, or κ-casein. κ-casein is responsible for making sure that casein, as a whole, turns into a Koosh ball. Without κ-casein, casein would look very different. Without κ-casein, casein would form a three dimensional mesh, held together by its interactions with calcium ions and by hydrophobic interactions between the other casein proteins. This is actually the exact type of protein network that we want to form during the early stages of cheese making.

Thankfully there is another protein whose sole job is to chop up κ-casein so that this network formation happens. Chymosin is a protein that is naturally found within the stomachs of young ruminant animals, like cows, goats, and sheep. The purpose of this enzyme is to curdle casein from the squishy Koosh ball shape to a more stable 3D network. The point of this enzyme is to give the nutrients found in milk a longer time to find their way into the animals during digestion. The stomachs of young cows literally turn milk into cheese to make digestion more efficient.

While cows evolved to turn milk into cheese, our evolution to cheese making was likely much more happenstance, although no less influenced by nature. In his book, *On Food and Cooking*, Harold McGee charts the progression of preserving milk from needing to drink it right away, to curdling and salting sour milk, to heating milk with pieces of calf stomach, accidentally co-opting a cow's digestion, to make cheese.[4]

For a long time, this is how cheese was made. Boil milk with a piece of stomach. And, the stomach could add both the acid and the chymosin required to make cheese. Eventually people figured out how to extract chymosin from a stomach. When this happens, it's not pure. That is, the stuff we get from the stomach has chymosin along with several other proteins. Altogether, this mixture is called rennet. In fact, humans knew about rennet and cheese making long before we had any idea that chymosin was the most important part. Today, when we talk about curdling milk, most people refer to rennet as opposed to chymosin.

Because the availability of calf stomachs limited the amount of cheese we could make (early times) and because many vegetarians require a diet free from the products of animal butchering (more recent times), humans have had lots of reasons to look for a non-ruminant source of rennet. There are several plants that produce enzymes that will curdle milk, most notably thistles.

CHEESE & CHARGES

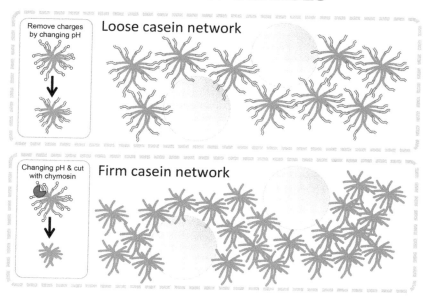

More recently, with the advent of genetic engineering, humans have been able to produce rennet on a large scale through the fermentation of yeast and bacteria.

For the final cheese produced, the network of proteins, and their relative strength, is defined by the curdling process (time and temperature and the type of mixing), the concentration of calcium, and the type of processing and aging that the cheese goes through. Calcium, for example, is partly responsible for the consistency of the cheese that gets made. Because calcium plays a role in bringing the casein particles together, it follows that the more calcium you add, the harder your cheese will be. This, in general, plays out in practice the way we might imagine. Cheddar and Colby cheeses are made in similar ways. One of the main differences is that the curds for Colby get washed in water to remove calcium. The removal of calcium is one of the reasons why Colby is not as hard as cheddar cheese.

I MELT WITH YOU

We return to our original problem of how to get cheddar to melt and blend into a cheese sauce. Any cheese's ability to melt and mix is defined by the sponge-like structure of curdled casein. The differences between each cheese can be understood within the finer details of this structure, that vary from one cheese to another.

Cake. Kitchen sponge. Natural sponge. All of these materials have the same generic skeleton: interwoven and interconnected fibers that surround empty pores and can soak up liquids. While they share a frame, each of these has a different firmness. You can cut cake with a fork and chew it. I would not suggest doing the same with a kitchen sponge, or a natural sponge, for obvious reasons other than for the purpose of testing their mechanical properties. Kitchen sponge is more flexible than natural sponge and is easier to tear.

The ability to melt the casein skeleton is affected by a number of properties. The calcium content is the first and, perhaps, the most obvious of these. More calcium means stronger interactions between casein and a cheese that is more difficult to melt into a liquid. Mozzarella is stringy because of its high calcium content; although it does melt, calcium ions hold strands of it together. Related to calcium, in this respect, is the acidity of the cheese. If the acidity in the milk is increased past a certain point, the protons start competing with calcium for interactions with the casein. High acidity, low calcium, cheese is more like the fresh cheese curds (fake Ricotta) that we talked about earlier. In this case, the casein framework will melt easier and won't be stringy.

Another factor that affects meltiness, is cheese's moisture content. Dry cheeses like Parmesan and Asiago and Pecorino, don't melt so much as turn crisp. The overall moisture content of cheese, then, keeps the casein skeleton from becoming too strong. Coupled to the dryness of the cheese, is its age. As you age cheese, casein interactions become stronger, making it more difficult to melt.

In sauce-making, the ability of cheese to melt is compounded by the fact that the milkfat is also going to melt. Milkfat is stored inside of tiny emulsified capsules that allow these particles of oil to hang out in water without pooling together. The milkfat in cheese will melt at just over 30 °C. If the casein architecture melts at near this temperature, the cheese will melt evenly into something that can be incorporated into a sauce. The implication of this observation is that as long as casein is in the same phase (solid or liquid) as the fat, it will keep the fat from pooling together. If the skeleton doesn't melt with the fat, the fat separates from the casein and you get a big oily mess.

Macaroni and Cheese

CHEESE MELTING

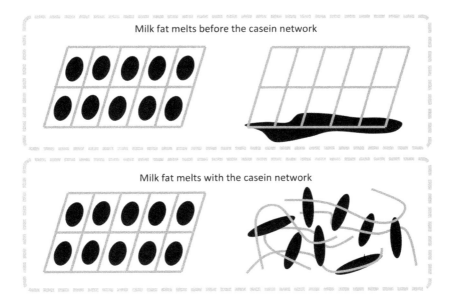

The latter of these two options is what happens with cheddar.

And, therein lies the chemical quandary of making a cheddar sauce. How do you incorporate cheddar, with all of its sharp, flavorful goodness into a sauce? Can you coax the fats, after they've melted and turned to liquid, to remain inside of the solid casein framework until that framework has a chance to melt?

All macaroni and cheese recipes seek to solve this chemical problem within the confines of the kitchen.

Kraft Macaroni & Cheese and other boxed mac and cheese products found a solution using a deconstructionist's approach. The sauce is made from the chemicals that are necessary for a certain texture, and the flavor is a mix of molecules that approximate cheddar. Of course an approximation will never live up to the richness and depth of the original. But, this chemical approach is certainly one way at this puzzle.

Historically, the best culinary answer for solving this problem was to stabilize melting cheddar with a roux, or a sauce that has been thickened with flour.

A typical recipe goes as follows. Melt 1 stick of butter in a heavy saucepan over low to medium heat and stir in 72 g

(6 tablespoons) of flour, making the roux. (The butter coats the flour and prevents it from clumping together due to gluten formation.) During this time, the butter and the flour brown a little and develop savory aromas. Add, with whisking, 980 grams (4 cups) of whole milk, bring to a boil, and keep at a simmer for 3 minutes. These are the magic moments, converting the roux to a Béchamel sauce, during which the flour goes from being tiny compact particles into swollen balloons that can thicken a sauce and stabilize the melting cheese. Slowly add 3 cups of cheddar cheese, a handful at a time. The thickness and viscosity of the sauce, before the cheese is added, prevents the oil from leaching out and pooling together. You literally slow the melted milkfat down to give the casein time to follow suit. It is important at this stage to not over-stir the sauce. Stirring helps the melted fat to find more melted fat and the solid casein to find other solid casein. Stirring is a recipe for a gritty, separated cheese sauce. Do so conservatively. Continue adding cheese in the same way until it is all incorporated.

For years, this type of recipe was the gold standard for cheddar sauces. The problem with roux-based cheddar sauces is the dilution of flavor and texture. In comparison to Guyère-based macaroni and cheese sauce, the roux sauce is thin and lacks the kick of the cheese in its un-diluted state.

Contrast the cheddar Béchamel macaroni and cheese with the following recipe for Gruyère macaroni and cheese from Mimosa, a French bistro in Los Angeles.[5] Mix 120 mL (1/2 cup) of Gruyère, 235 grams (1 cup) of cream, 245 grams (1 cup) of milk, sliced prosciutto, 3 tablespoons of nutmeg followed by the addition of cooked macaroni. Place in a baking dish and top with 1 cup of Gruyère. Bake until cheese melts. Gruyere can be thickened with a roux, but it doesn't need to be. It melts fine all on its own. The cream and milk used in this recipe add enough liquid to get the consistency of a sauce that will coat the pasta. Otherwise, it's unadulterated cheese.

The chemical games that we play while making macaroni and cheese come in all sorts of forms. The destructionist powders are one way at it. Roux based sauces are another way at it. Yet each of these has left me with a desire for more cheddar taste in my macaroni and cheese.

There are two more recent entries into the mac and cheese world that I have really enjoyed. These recipes use typical pantry items but aren't afraid to tout the chemical nature of why they work so well.

The first recipe is from *Ideas in Food* by Aki Kamozawa and H. Alexander Talbot and is frighteningly easy.[6] In a pot over medium heat, add 42 grams (3 tablespoons) of butter and one 355 mL (12-ounce can) of evaporated milk. Throw in some extra seasoning for flavor; they suggest cayenne pepper and salt. When the milk begins to steam, add 280 grams (10 ounces) of grated cheddar cheese and 280 grams (10 ounces) of grated pepper Jack cheese. Add by the handful and allow each addition to slowly melt before adding more. Avoid over stirring.

The key to cheese stabilization in this recipe is found in the evaporated milk. On its own, milk contains a number of emulsifiers to keep the milkfat from separating. In its evaporated form, the emulsifiers are found at a higher concentration. Smaller package, just as much stabilization. On top of these, evaporated milk has another additive, carrageenan. Extracted from seaweeds, carrageenan is a polysaccharide that has been used alongside dairy in a number of recipes as a thickener. Carrageenan is also used to prolong the shelf-life of foods like evaporated milk. One can make a comparison, chemically, between a Béchamel sauce and carrageenan-stabilized evaporated milk. In both cases a hydrated polysaccharide is thickening a dairy-based liquid. Certainly there are flavor differences between the two. But from a functional point of view (they can both keep cheese sauces from separating by exploiting polysaccharides) they are similar.

The point of Kamozawa and Talbot's recipe is to make things as easy as possible. Instead of boiling the pasta, they just steep it in room temperature water for an hour. They take a shortcut around making a roux, which can be intimidating for some home chefs, by using evaporated milk. And, I must say, that I'm a sucker for their flavor combination. The pepper jack cheese and the cayenne pepper add a wonderful kick. When I make this recipe at home, I have to forgo the spice as my kids don't have the tolerance that their father has. However, when I make it during my chemistry of cooking class, I usually make one milder batch and one batch with a little extra kick. Aside from its ease of preparation, it's the one aspect of this macaroni and cheese that my students always comment on.

The last macaroni and cheese recipe that I am going to discuss, and it may be my favorite, is the recipe from Kenji Lopez-Alt's book *The Food Lab*.[7] As a word of caution, I'm going to describe each step and what's going on with the science. However, if I start prepping ingredients when I start the water boiling, I can finish this recipe 5 minutes after the pasta is finished cooking. So, my description may make it seem like it takes longer than it actually does.

Grate 1 pound of extra sharp cheddar cheese and 1/2 pound of American or Jack cheese. Toss this cheese with 1 tablespoon of cornstarch. The cornstarch makes sure that the cheese melts evenly. It also thickens the sauce as it cooks. In a separate bowl, whisk together one 12-ounce can of evaporated milk, 2 large eggs, and seasoning (Lopez-Alt suggests hot sauce and ground mustard, I use Dijon mustard and cayenne pepper). The evaporated milk acts the same in this recipe as it does in the *Ideas in Food* recipe. The eggs I'll get to in a bit. But, they are the real secret of the whole thing. When the pasta is cooked and drained, toss it with the butter until the butter is melted and coats the pasta. Add the evaporated milk and egg mixture and mix well. Add the cheese and continue stirring until the cheese melts. Eat and enjoy.

You'll notice one big difference between this recipe and the other two recipes (other than ingredients). Lopez-Alt's recipe allows for stirring the cheese in with the pasta and the sauce while it melts. The other recipes call for delicate and careful stirring of the cheese to make sure there is no separation of the milkfat from the casein. This difference is a result of several ingredients, but is primarily facilitated by the eggs. The cornstarch, as mentioned earlier, helps the cheese to melt more evenly. It coats each strand of cheese and helps to separate individual strands in a way that helps to prevent the pooling of milkfat from multiple cheese shreds. Again, the evaporated milk works as previously described.

The eggs really steal the show in this recipe. While the cornstarch and carrageenan physically slow the motion of the fats to keep them incorporated within the casein framework, the eggs, in a sense, glue the oil to the skeleton. Egg yolks contain the emulsifier lecithin, which has one end that interacts strongly with oils and another end that interacts strongly with hydrophilic molecules. It's like having double sided tape, with each side able to stick to two completely different things. The eggs also make a custard that works to coat each piece of pasta.

The result is really satisfying.

Lopez-Alt's recipe is a chemistry-heavy approach that makes use of some very versatile, everyday pantry ingredients. And it yields a sauce with the requisite cheddar flavor to satisfy a picky guy like me.

REFERENCES

1. K. Brix and W. Stöcker, *Proteases: Structure and Function*, Springer, Berlin, 2013.
2. R. Carroll, *Ricotta: Everything You Ever Wanted to Know About This Simple Cheese*, [document on the Internet], Cheesemaking.com; [cited 2016 July 5], available from: http://www.cheesemaking.com/store/pg/217-Ricotta.html.
3. S. Spangler, *Naked Eggs – SICK Science!*, [document on the Internet], Stevespanglerscience.com; [cited 2016 July 5], available from: http://www.stevespanglerscience.com/lab/experiments/naked-egg-experiment/.
4. H. McGee, *On Food and Cooking*, Scribner, New York City, 2004.
5. *Macaroni and Cheese with Prosciutto*, [document on the internet], Epicurious, [cited 2016 July 5], available from: http://www.epicurious.com/recipes/food/views/macaroni-and-cheese-with-prosciutto-104829.
6. A. Kamozawa and H. A. Talbot, *Ideas in Food: Great Recipes and Why They Work*, Clarkson Potter, New York City, 2012.
7. K. Lopez-Alt, *The Food Lab*, W. W. Norton, New York City, 2015.

CHAPTER 7

Bread

Sometimes, one of the most difficult things to do while cooking is to change the recipe. Part of this is a mental block. We're given a recipe, the recipe is supposed to work, changing the recipe will result in something that doesn't work. We don't trust ourselves or our experiences in the kitchen. My wife is more of a strict recipe follower than I am. She's one of the best cooks that I know. But, many times, she trusts a recipe author's instincts more than her own, which are hard earned and more developed than she gives herself credit for.

There have been more than a few times when we've been making dinner together and I'll have been poking and prodding and tasting the food that we've been making. I'll add some seasoning or alter the heat or flip a piece of meat inside of a casserole dish before I'm supposed to. She'll get mildly annoyed with me when I do this. What really sets her off on me is when I'll pull a roast out of the oven before the prescribed amount of time is up. Her ire is most apparent when she is the one of us who started the dish in the first place.

My perception of her reliance on recipes is colored by the fact that I am overly fidgety in the kitchen. I do like to poke and prod and tinker, probably too much. It's not that I know what I'm doing. I poke and prod and tinker because, in most cases, I don't know what I'm doing. I'm trying to understand.

Chemistry in Your Kitchen
By Matthew Hartings
© Matthew Hartings, 2017
Published by the Royal Society of Chemistry, www.rsc.org

Some might think that scientists come out of the womb naturally curious, probing the world with every trip to a playground or morning spent in a classroom. In fact, there are many in the world of science education who, wrongly I think, claim that all children are born scientists. All children are born loving to play and experience, and they are really good at learning by doing. It seems to me that there is a severe deficit in the amount of free playtime we give our children during their schooling. The inability to dictate any part of their own schedule and have a little freedom in what and how they learn can have some really undesirable effects, especially in terms of student confidence.

When I was going through my PhD training at Northwestern I found myself struggling with confidence issues. I don't think that this was or is abnormal. Many people question their abilities during graduate school. There is a term, "Imposter Syndrome," that describes people who feel that they are going to be found out as frauds even when their successes are apparent.[1] Because our work as scientists takes us into a place where unknowns abound and prescribed actions are few, many of us feel like imposters well after our schooling and training is over.

While there is no shortage of scientists who have stories like this, I hope that sharing mine can give you confidence to experiment in the kitchen and trust what you learn.

I started graduate school working for a chemist named Martin Jarrold. Martin is well known for building big instruments that study the shapes of really little things. I was working on a project that used these instruments to probe the size of tiny proteins. These machines were really cool. They could strip away all of the water from around the protein, shoot it into a chamber, and bombard the protein with argon atoms. Using our own stopwatches, we could measure the time it took for the protein to get from one side of the chamber to another. We used this measured time to figure out the shape of the protein, determining what kind of origami it had folded itself into.

I really enjoyed this research and found myself thriving within it. We were publishing papers and designing new experiments. The early successes, which, when I look back on them, had as much to do with how the experiments were set up (by my advisor and a senior colleague), as my hard work put into completing them.

At the end of my second year, my advisor took a new job at another university. I had a choice, to either follow him to that university or to start over with another advisor at Northwestern. Long story short: I decided to stay at Northwestern. Erika and I married the same week that the lab moved. We really loved Chicago. She loved her job there and her friends and colleagues. Picking up and moving just didn't make sense to us.

So, I started a new project. I signed on to do some research, understanding how electrons move through proteins, which is a key process in the way that our bodies turn oxygen into energy. To me, the project was new, the science was new, the tasks were new, and my approach had to be new. Working as a team, we were successful. But I never felt that my contributions were as critical or as insightful as they needed to be.

One of my advisors was Mark Ratner. Outside of my family, I have never known a better person. He was welcoming of my wife and me in a way we never really deserved. He is also one of the most brilliant minds in science. He has spent his career studying how electrons move and bounce from one place to another. Mark's greatest quality, though, is he can instinctively make you feel smarter than you actually are. I have seen him give research seminars at international conferences. Talking about some rather esoteric math and even more esoteric chemistry, there are maybe 5 people in the world who actually know what he's talking about. But, when Mark gives a presentation, every single person in the room, in that moment, understands his research perfectly well.

I remember struggling with my own research and going to talk to Mark about it. We chatted about focusing on what was important, what would make the research sing for the people who cared about what we were doing. I remember him saying, "I've got some ideas for how this might work." And that was it. I wanted more. Mark had faith and trust in me, and I was unsure. I suffered some anxiety in that I didn't really know how to start with what we had discussed. Because the project was new to me, I felt like, however I chose to start, was going to be wrong. That feeling was crippling to me.

Throughout the rest of my PhD, I was productive. But, with this challenge that Mark and I had come up with, I never really saw it through the way I wanted to. My doubts were founded. Science demands that we try things that aren't correct. Often you don't know what is correct, or not until after you've tried. The project that he and I talked about was so important to me that I didn't want to err. I didn't want to let him down. In the end, I never did pull through on this. My failure in this was never an issue to Mark. I internalized the success and failure of this research so much that I never truly allowed myself to start.

I've realized much later that I wanted something prescribed, something safe, something that was going to give results that I could publish. I think that the fact that I cared about this project so much made me push it away because I couldn't stand to see it fail or stand to see myself fail at it.

Now that I've become an educator, I see my own shortcomings much more clearly. I want for my students to embrace failure. I want them to run headlong into failure. I want them to fail so much that they stop fearing it. I want them to fail so much that they can recognize its early stages, remain calm, and change course. I want them to disregard failure so much that when they start working on their own projects they aren't crippled the way I was.

For the chemistry students that I teach, this means getting rid of traditional, cookie cutter labs that are meant to work no matter what. Science has never been about doing what is known and expected. The purpose of doing science is to understand things that no one else has had the foresight or the ability to understand before you came along. At the university level, I want to indoctrinate our youngest students into a working research project. In traditional teaching labs, experimental failure usually means that your execution was off and that should correspond to poor grades. This is not how it should be. I want my students to experience and cause failure, so that they know how to respond when a project goes wrong.

I want to give our senior-level students a slightly different experience. I want them to be able to define what area of the unknown they care about. And, I want them to be able to figure out how they scientifically probe that area. I want them to be able to build strategies and protocols that are robust to failure. That is, if they have experiments that fail, and they will, they need to have backup plans and escape routes for moving their projects away from something that isn't working.

There is very little in this world that we know will work every time. Chief among these is the pursuit of science, where we should expect failure because research actively seeks the unknown. Whether my students go into science, or medicine, or government, or business, or communication, or education, they will be working in areas where there is uncertainty. I want more than anything for them to have the skills and confidence to navigate those murky waters.

I do my best to instill these same attitudes and skills in the students who come through my chemistry of cooking course as well. These students are usually first or second-years, they are certainly not science majors, and they are typically high-achievers with high expectations for themselves. I spend my entire semester surrounding them with failure. They certainly don't fail the class; that is not my goal. In fact, the grades I hand out in this class are generally very high.

I want to provide them with a safe place to fail. I want them to be in situations where failure in replicating a recipe isn't reflected in their grade. I want them to be able to see me fail as well, which they do... often. My class is some strange mixture between a science lesson, cooking show, and performance art. I mess up all the time. It's impossible not to.

I make something at the start of every class. Caramel sauce, steak, and ice cream, are all on my menu throughout the semester. There have been times in class where I have scorched a pile of sugar when adding it to a pan that was much hotter than it needed to be. I left trails of smoke as I sprinted out of the classroom, removing the pan before the fire alarm could go off. I have failed, more times than I care to count, at making mayonnaise using an immersion blender. Using a prep that was supposed to be foolproof, my egg and oil mixture would always break, forever lost as a runny and goopy mixture, unable to actually become mayo. I have also made macaroni and cheese that is so gritty that you could use it as sand paper.

One of the places I know that failure, or students perceiving that they have failed, will show up, is when I start the section that focuses on reverse engineering a recipe to try to figure out what role different ingredients play in a dish and using that understanding to substitute ingredients while not changing the dish.

In the classroom, some of my students got to see failure on full display when I tried to make vegan caramel sauce. Caramel sauce is my favorite thing to make all semester long. First off, it's delicious; who can resist apple slices dipped in caramel sauce. It is a wonderful recipe that demonstrates how you can see, smell, and taste chemical transformations as they happen in a kitchen. [Heat

200 grams (1 cup) of sugar in a pan. Allow most of the sugar to melt before stirring. After the liquid browns to a tan color, add 115 grams (1/2 of a cup) of heavy cream little by little, stirring between each addition.] The recipe itself is quick to make and can cool enough for students to try it out before the end of class. I even have some time built-into the schedule so that I can make it multiple times, should I fail in making it correctly the first (or second or third) time. Finally, I get to show the students, who might be intimidated by chemistry but are not intimidated by making caramel sauce, that the chemistry behind caramel is incredibly complex. The chemistry of caramel intimidates and perplexes most professional chemists. The students get to see that they can perform and master complex chemistry in the kitchen.

After hitting my stride teaching this class, I realized that many of the dishes that I cook reflect both my preferences for food, as well as my heritage. (Admittedly, this book does the same thing. The food we enjoy and that most comforts us is more related to the culture we have grown up in than any particular chemicals that they contain.) Wanting to make my class more inclusive for people who keep vegan diets or who cannot eat dairy, I thought that making caramel sauce without heavy cream would be an easy enough way to start being more welcoming to students with different backgrounds and needs than me.

I also thought that making a vegan caramel sauce would be an excellent way to demonstrate how to substitute ingredients to make a recipe fit your needs. The easiest way to see what role an ingredient plays is by removing it from the recipe. Carmelizing sugar on its own will result in a very tasty film of hard candy. There is no sauce without the heavy cream. In addition to this, if you substitute the heavy cream for half and half or whole milk, the sauce doesn't come together at all. You get kind of a mess. This observation told me that a higher fat content is needed to make a good caramel sauce. So, the first time I tried to make a vegan caramel sauce in class I used coconut oil as a substitute for heavy cream. Coconut oil has become a popular staple in many pantries even though it has high levels of unsaturated fat. Anyway, I tried to use coconut oil and it failed miserably. The oil and the caramel wouldn't mix.

I remember trying variations of this recipe multiple times throughout the semester, all of them going wrong. I eventually settled on using a recipe for a vegan-like caramel sauce. [Add 392 grams (1 14-ounce can) of coconut milk and 165 grams (3/4 of a cup) of brown sugar to a pot and heat until the sauce has thickened to your desired consistency.][2] In this recipe, multiple substitutions are made. Coconut milk provides the basis for the sauce and the creamy mouthfeel that we expect for a caramel sauce. The brown sugar replaces the flavor of the caramelized sugar and helps to thicken the sauce into more of a syrup. For my current students, I don't try to make the coconut oil recipe any more. (There are plenty of other occasions where I will screw up over the remainder of the semester.) While I'm making the coconut milk and brown sugar recipe, I do recount my mishaps while showing them what you need to think about while substituting ingredients and how to navigate away from bad recipes when things start to go wrong. Most importantly, I tell them, "Its just food. So what if you mess up? You can just try again. And if that goes wrong, order a pizza."

We set our students up to try recipe substitutions on their own in the laboratory portion of the class. One of my favorite activities is the cookie lab. Students start off making the shortbread cookie recipe from Michael Ruhlman's book *Ratio* (1 part sugar, 2 parts butter, 3 parts flour).[3] As detailed in his book, minor changes in these ratios and simple substitutions can result in completely different cookies. Increasing the butter will give a thinner cookie. Swap granulated sugar for brown sugar, the cookie will be moister and have more pronounced flavor. In the lab, we ask the students to make any substitutions or ratio changes that they like. They bake up their batter and note the differences between the standard recipe and their recipe. Then, they come together as a class to discuss their changes and the effects on the cookies. Finally, the students are tasked with making a cookie with different properties (chewiness, flatness, coloration, *etc.*) that are defined by the instructor. The students have to use the accumulated class data from their own recipes to figure out how to make the cookies that the instructor wants. It is a really fun lab, and it is an excellent introduction to navigating from failure to success when changing a recipe.

CITIZEN SCIENTISTS

One of the mantras I repeat while teaching this class is that anyone can do chemistry. Much like Chef Gousteau in *Ratatouille*, saying, "Anyone can cook," I hold fast to the notion that good science can come from anywhere. Taking this a step further, science, as a whole, is better when there are more people involved in the scientific process.

I love kitchen chemistry because I always end up finding people from disparate backgrounds (scientists, professional chefs, food writers, and people who are just dabbling with cooking out of curiosity) doing some really impressive chemistry. I see chemistry everywhere. Erika, at times, thinks that this is an affliction. But, I am conditioned by my training pull chemistry from the depths of the everyday.

There are some very impressive kitchen chemists out there. Home brewers, people dabbling with modernist cuisine, and back-yard pit masters have developed and perfected some really impressive chemical tricks. But, the group of people whose work I am most fascinated by includes those who have tried to develop gluten-free bread recipes.

I have a love affair with gluten. My love of gluten is apparent within the pages of this book. With chapters on pancakes and bread and pizza and pie, I have dedicated lots of space and thought to foods that feature gluten. And, to me, the food that best captures the magic of gluten, is bread.

I would be happy eating freshly baked bread for one of my meals every day. I might have inherited this trait from my mother. We didn't go out to eat a lot when I was young. But, from the few times that I remember going for a nice meal, I vividly see my mom not holding back when the waiter would bring the breadbasket around. To this day, she has to watch herself when she goes out for dinner so that she doesn't fill up on bread before her dinner arrives. It was always such a luxury to have a fresh loaf of good bread at home. I still feel this way even though I bake bread at home once a week during the cooler months. It seems that this trait has been passed on to my children as well. For her past several birthdays, my daughter Claire has requested bread and cheese and fruit for her special meal. I am all too happy to oblige.

The chewy bite in a loaf of bread, its ability to trap little pockets of air as the dough expands, and the browning in the crust, all come about because of the presence of gluten. Gluten is made when two proteins from wheat, glutenin and gliadin, come together to form a web of interactions. As such, gluten is defined as much by the layout of this web as it is by the identity and properties of the proteins that make up that web. Glutenin and gliadin are secreted, along with starches (amylose and hemiamylose), by different members of the grass family (wheat, barley, rye, *etc.*). Other dietary starches (cornstarch, tapioca starch, and potato starch, to name but a few) don't have any significant number of proteins associated with them. They certainly do not contain these gluten-making proteins.

The effects of gluten are rather obvious on quick inspection. Flour and cornstarch are rather similar to one another. Flour has more color and feels a little softer. Individual crystals of cornstarch appear to be smaller than their flour counterparts. Both are mostly made up of polysaccharides. Cornstarch is solely made of polysaccharides. 90 percent of flour is made up of those same carbohydrates. Both are used in nature as stored energy sources that get packed into, and for the benefit of, seeds as they germinate and grow into saplings.

Adding water to either cornstarch or flour will immediately reveal the differences between having and not having glutenin and gliadin. If you are trying to maintain a solid structure with cornstarch when adding water, the best you can hope for is a paste. While you can't make bread with it, this paste is interesting on its own. It is pourable and will slosh around in a container. However, if you exert a swift and sudden force upon it (by smacking it, for instance) it will take on a solid structure. We call this type of substance a non-Newtonian material. Some people refer to cornstarch paste as Oobleck, in reference to the green goop described in Dr Seuss's *Bartholomew and the Oobleck*.[5] Oobleck is

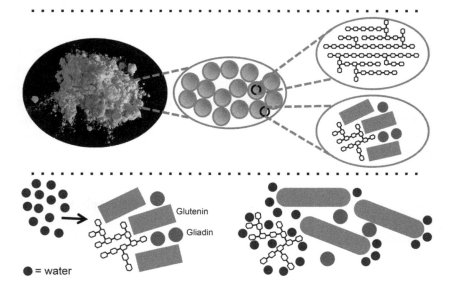

FLOUR

a favorite of people doing science demonstrations. If you place a speaker under a container of cornstarch paste, the sound waves traveling through it will cause it to rhythmically change from liquid (when no sound is going through) to solid, holding the shape of the speaker until the sound dissipates.

The addition of water to flour couldn't be any more different. One of the first things I notice when I mix flour and water is how sticky the mess becomes. Getting the stuff off of my hands is always more work than I expect it will be. It also has a sort of shaggy look to it. Contrast this with cornstarch paste, which seems to have a sheen on it from the way it evenly reflects light. Gluten, these observations quickly show, is the glue that holds flour and water together in a dough. And, a few simple experiments and observations can help us to better understand how that glue works.

Flour does not get sticky until you add water. (This is obvious but is a necessary logical observation to make.) Flour does not become sticky when you add oil. There is something about glutenin and gliadin that causes them to be disrupted from the starch crystals in flour upon the addition of water. After you make dough, you can continue to wash it with water until you are left with a smaller, solid mass of goo. This bunch of goo actually represents all of the gluten that forms when you first make the dough.

So, water initiates gluten formation, but it cannot break gluten down and dissolve it. Oil, on the other hand, prevents gluten formation. We apply this science every time we make a roux or use flour to thicken gravy. In both cases, oils are used to coat the flour and prevent it from making a dough when it is added to a water-based liquid. Moreover, oil is used to weaken the gluten in many baked goods. Shortbreads are made from batters that contain butter or some other sort of oil. Shortbreads are often soft and are sometimes flaky. Gluten development in these baked goods is altered by the addition of oil.

From these observations, we can build a chemical description to explain how gluten formation works. Glutenin and gliadin are held onto the starch with hydrophilic interactions. This is no surprise as the starch itself is replete with –OH groups, which are themselves prone to hydrophilic interactions. When water is added, these molecules displace the proteins from the starch. From here, hydrophobic interactions drive the intermingling of the proteins.

When gluten first forms, it does so haphazardly. The initially shaggy and sticky dough is an indication of this. The fact that the dough is so sticky, is an indication that the proteins are looking

Chapter 7
Kneading & Gluten

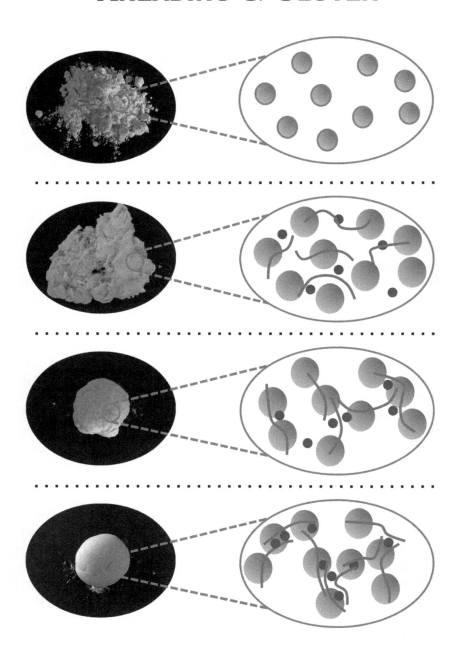

for something to interact with. It is only through kneading of the dough, that full gluten development can take place.

Kneading itself is an incredible act. It is at once a tactile, a kinetic, and a chemical act. Taking the sticky, unformed dough, on a lightly floured surface, it is pushed, pulled, flipped and turned. Kneading is a very rhythmic process. The dough is flattened and then gets stretched forward, folded over, and rotated a quarter turn. Stretch. Fold. Rotate. Stretch. Fold. Rotate. The acts are simple. But their consistency and repetition transforms the dough in ways that are both obvious and unseen. The rough and undisciplined dough is transformed into a ball that is smooth, reflective, taut, and firm. That the dough slides off the hand afterwards, is an indication that the proteins are now all neatly interacting with one another. That the dough is reflective, is an indication that the neatly interacting proteins are also neatly aligned.

In *On Food and Cooking*, Harold McGee shares several images of what developed gluten looks like when magnified by a powerful microscope.[6] The linear arrangement of protein fibers is reminiscent of the kinds of fibers that form and go on to make plaques in people who suffer from several aging-related diseases.[7] Alzheimer's, Parkinson's, and Huntington's diseases are all examples of these. It is no coincidence that the same kinds of hydrophobic forces drive the formation of both gluten and the disease-related plaques.

That some similarity exists does not make gluten inherently dangerous or toxic. Certainly, there are some people who suffer from celiac disease. But this has to do more with an inability to digest gliadin and the autoimmune response that goes along with that.

I find such pleasure in the truth that kneading aligns proteins. That the motion of our hands on the dough translates down to the molecular level, causing the proteins to twist, turn, and stretch with every movement, puts me in awe of both our own abilities and the properties of the molecules. It also drives it home for me that energy comes in many forms. We often think of energy as heat and temperature; even chemists and physicists subscribe to these misperceptions. But energy also comes in the form of motion and kinetics. My wonderment at the translation of manhandling to molecular motion is matched in equal parts for the way that molecular alignment is translated up to a scale that we can see and feel. The proteins shape-shift from a random mesh to tightly aligned fibers and we see the dough going from rough and shaggy to smooth and silky. The way we experience food is always controlled by what's going on at a molecular level, even when we can never hope to literally see them on our own.

I must admit that I have a hard time figuring out when I've finished kneading when I make bread. One of the reasons why I prefer to make Jim Lahey and Mark Bittman's no-knead recipe is that the long rise time (16 hours) takes care to make sure that happens.[8] But, for the intrepid bread makers out there, there are a few things to look for in dough with fully developed gluten, outside of the dough being smooth and silky and shiny. It should stretch far enough and thin enough so that light can shine through it. Dough's ability to do this is usually attributed to gliadin, which acts like little ball bearings. The dough should also be elastic; it should collapse back onto itself after it's been stretched. Glutentin gets credited with this property. Specifically, there are sulfur atoms on glutenin that are always in search of other sulfur atoms. As the individual glutenin proteins pass by one another, their sulfur atoms grab, hold tight and ensure that the dough remains strong, even when you are pulling on it.

Kneading grants the dough with all sorts of other important properties. Kneading and forming the dough creates all of the miniscule creases and cracks and spots that will eventually become the crumb, the pattern of air pockets that run through a loaf. While most of those bubbles are made up of carbon dioxide with a smattering of water vapor, it is important that oxygen gets mixed into the dough as well. The sulfur atoms that I just mentioned, aren't quite ready to grab onto each other without a little priming from oxygen. Oxygen is also important for turning some of the carotenoid coloring compounds from flour that we see with our eyes, into flavors that we enjoy with our mouths. As with all things, this oxidative chemistry is a delicate balance of "need to do it" with "better not over do it." Over-mixing and over-kneading are legitimate concerns when you are using stand mixers.

The magic of gluten is that it plays so many roles in bread. It traps bubbles. It defines the firmness of the bite needed for chewing. (This is seriously important. The difference between bread and cake is partly due to the low gluten flour used to make cake batter.) Gluten holds the dough together. It gives the dough elasticity. It allows the dough to stretch properly. The gluten is also partly responsible for the Maillard browning, which gives the bread extra flavor and the crunch and crackle of the crust.

Any gluten-free bread recipe that wants to actually recreate the texture of bread will have to find ways to compensate for all of these effects. This is no simple task. However, for anyone who suffers from celiac disease or has a gluten sensitivity, the quest for good bread is well worth it.

Shauna James Ahern is a writer and cookbook author. On her blog *Gluten-Free Girl*, she chronicled how she was diagnosed with celiac disease and how, after this diagnosis, her life and her cooking changed.[9] Like me, she loved baking and baked goods and, perhaps, liked bread a little bit too much. Her blogging chronicles many of her efforts to create a new normal in her life, where she could enjoy the same foods as she did before her diagnosis. Her writing, done as a team effort with her husband, Daniel Ahern, resulted in the publication of several excellent cookbooks with gluten-free recipes, most notably *gluten-free girl*.[10] Certainly the increased number of celiac and gluten-sensitivity diagnoses is part of why their books are so popular. But, Shauna's writing is so infectious and her recipe descriptions are so approachable and engaging, that the books would have been popular anyway.

What is particularly attractive to me about her writing is the way that she describes her recipe development: her goals, her thought process, her frustrations, and her successes. In many of her recipes, there is a resemblance to America's Test Kitchen, but with a little more warmth and humanity and personal flair. And through it all, she does some really amazing science. It is obvious that she and her husband keep good notes of ingredient combinations that they try. They identify and refine different variables. And, through this process, they converge on a recipe. Kitchen chemistry at its best! One recipe of hers stands out in particular.

Like any bread lover who has just found out that they have celiac disease, Shauna needed to find a passable bread recipe. And the biggest problem that she faced was figuring out how to replace the gluten.

Many gluten free recipes call for the use of xanthan gum. Xanthan is a hydrocolloid that gets used in many gluten free recipes. Thomas Keller, arguably the best chef in America, markets a gluten-free flour recipe called Cup4Cup.[11] Its ingredients include brown rice flour, white rice flour, ground golden flaxseed, rice bran, and xanthan gum. Aki Kamozawa and Alex Talbot describe their process for developing gluten free flour on their blog *Ideas in Food*. Their mixture, which they call, "What IiF Flour," contains cornstarch, tapioca starch, white rice flour, brown rice flour, milk-powder, and xanthan gum. They have also written a book, *Gluten-Free Flour Power*, about gluten-free baking that highlights how to use and modify their flour when baking different dishes.[12]

Xanthan gum is known in molecular gastronomy circles, as a food ingredient that helps to stabilize foams. Xanthan is naturally produced by a bacteria, *Xanthomonas capestris*. You can actually see these bacteria in action at home. Any lettuce that has been left in the fridge for too long gets a little slimy. That slime is xanthan gum. On its own, xanthan gum isn't bad for you. But, it is an indication that your lettuce is starting to go. It is funny how these different chemicals find their way into the food world. But, because of its unique talents, xanthan has become a staple for many chefs and adventurous home cooks, due to its ability to stabilize and thicken sauces and, mostly, for its ability to trap gas. Add a little xanthan to some juice, pump in some air (either with an immersion blender or a whipping canister), and you can make some delightfully light and flavorful foams. The foaming ability of xanthan is what makes it a natural ingredient in gluten-free recipes. It also provides a little chewiness to whatever it's added to.

But, for Ahern's money, xanthan doesn't cut it. She notes that she does have a slight digestive sensitivity to xanthan and guar gum, another commonly used ingredient. But, her biggest problem with them is that they make her baked goods a little too... gummy. She gets around the use of xanthan and substitutes for gluten by making a slurry of ground flaxseed meal and chia seeds, in hot water. Hot water is needed for this step as it helps to extract and hydrate the mucilage contained in these materials. Mucilage is a complex mixture of biological molecules (proteins and polysaccharides) that are used to help keep things, like seeds, safe from bacterial infection.

Along with these ingredients she uses oat flour, almond flour, teff flour, potato starch, arrowroot powder, buckwheat flour, milk powder, sugar, salt eggs, apple cider vinegar, water and yeast. All of these, in some way, make up for not having gluten. She freely admits that there is still something missing, writing, "Now let me remind you of this: there is no substitute for gluten. Gliadin and glutenin combine forces to create elastic binding in doughs that I will never achieve, no matter how many different flours and slurries and starches I try. Those of you who can eat gluten? Please start baking bread. You have it so easy."[13] For her recipe, some of the flours and the vinegar help to add flavor. The egg and milk powder help to hold the dough together.

You'll find milk powder in the gluten-free flour ingredients from Keller as well as those of Kamozawa and Talbot. Remember, the glutenin unfolds and realigns itself with some prodding from mechanical energy (kneading). The proteins in milk (and in eggs) also unfold when mixed rapidly; milk or egg based foams are both produced by whisking. While they will never align themselves or hold air bubbles in the same way that gluten does, they can provide some of the texture that goes missing when gluten is removed from the equation.

What I admire most about Ahern's description of this recipe, though, is the way she talks about her perceptions and how they tricked her into thinking that one of their recipes was bad.[10] See, she had baked bread before. And, having experience with that, she expected her dough to have a certain consistency. The first time she made this dough, however, it looked really dry and crumbly. So, she added more water to it. After baking, the bread came out really soggy and over-damp. Going back to the recipe, as it was written, she gave it another try. Even though the dough was grainier than she had expected, the bread came out with the consistency that she wanted in her final product.

For my classes, I took the liberty of analyzing her recipe and use Michael Ruhlman's bread recipe from *Ratio* as a measuring stick.[3] Ruhlman's recipe calls for 500 grams of flour and 300 grams of water, along with 1 teaspoon of yeast and 2 teaspoons of salt. The ratio of solids to liquids is 1.67. Analyzing Ahern's recipe, the ratio comes out to be 1.56. The hydration levels of each are similar. Even though her dough looked dry, adding more water increased the dough's hydration. I love this example because it shows a scientist's attention to detail. She noticed that something looked off at the start. She did a controlled experiment to see if she was right. She found out that her instincts were wrong and kept the original recipe.

So, even when we have good kitchen instincts, a mind for the scientific process can help.

Changing ingredients and recipes on the fly can be nerve racking. But, in comparison to laboratory chemistry, you get to eat your experiments when you're finished. And, if it doesn't turn out how you wanted, it's really not a big deal. Order some pizza and try again another day!

REFERENCES

1. *Impostor Syndrome*, [document on the Internet], Wikipedia; [cited 2016 July 5], available from: https://en.wikipedia.org/wiki/Impostor_syndrome.
2. *Coconut Dulce de Leche*, [document on the Internet], Epicurious; [cited 2016 July 5], http://www.epicurious.com/recipes/food/views/coconut-dulce-de-leche-240759.
3. M. Ruhlman, *Ratio: The Simple Codes Behind the Craft of Everyday Cooking*, Scribner, New York City, 2009.
4. *Ratatouille*, Hollywood, Pixar, 2007, Film, directors Brad Bird and Jan Pnkava.
5. D. Seuss, *Bartholomew and the Ooobleck*, Random House Books for Young Readers, New York City, 1949.
6. H. McGee, *On Food and Cooking*, Scribner, New York City, 2004.
7. J. R. Harris, *Protein Aggregation and Fibrillogenesis in Cerebral and Systemic Amyloid Disease (Subcellular Biochemistry)*, Springer, Berlin, 2012.
8. M. Bittman, *The Secret of Great Bread: Let Time Do the Work*, New York Times, 2006.
9. S. J. Ahern and D. Ahern, *Gluten-Free Girl*, [document on the Internet], Gluten-Free Girl; [cited 2016 July 15], available from: https://glutenfreegirl.com.
10. S. J. Ahern, *Gluten-Free Girl: How I Found the Food that Loves Me Back ... and How You Can Too*, Wiley, Hoboken, New Jersey, 2009.
11. T. Keller, *Cup4Cup*, [document on the Internet], Cup4Cup.com; [cited 2016 July 5], available from: http://www.cup4cup.com.
12. A. Kamozawa and H. A. Talbot, *Gluten-Free Flour Power: Bringing Your Favorite Foods Back to the Table*, W. W. Norton, New York City, 2015.
13. S. J. Ahern, *Gluten-Free Multigrain Bread*, [document on the Internet], Ruhlman.com; 2011 February 14 [cited 2016 July 5], available from: http://ruhlman.com/2011/02/gluten-free-multigrain-bread-recip/.

CHAPTER 8

Vinaigrette

How do you mix oil and water?

The easy answer to this question is that you don't.

All of us inherently know that oil and water just don't play well together. The chromium-spectrum array of colors that show up in parking lots after a rain are only there because a thin layer of motor oil sits atop a puddle of water.

Nowhere are we more cognizant of the oppositional nature of oil and water than in our food. This is readily apparent at the grocery store while looking at any bottle of vinaigrette, the layer of oil riding on top of a base layer of acid. It is no exaggeration to say that much of cooking involves finding ways to integrate oil and water into a single liquid and then devising means to keep that liquid from separating.

A typical recipe for vinaigrette that you'll find most anywhere (on-line or in cookbooks) requires a flavorful acid (citrus juice or some flavor of vinegar are often used) and olive oil. Different seasonings can be added and the ratio of oil to water can be changed to enhance the flavor, but this recipe doesn't change too much. In nearly all of these recipes, the acid is added to a small bowl and olive oil is slowly added with copious whisking. You can quickly add this vinaigrette, before it separates, to dress a salad. The beauty of vinaigrettes lies in that they are easily and quickly

---

Chemistry in Your Kitchen
By Matthew Hartings
© Matthew Hartings, 2017
Published by the Royal Society of Chemistry, www.rsc.org

made at home with ingredients that most people have on-hand. And, after you get a feel for how much of it you use per sitting, you can make only as much as you'll need for a single meal. There's no waste, no storing bottles of dressing in the refrigerator, and you have the ability to quickly conjure up a dressing to complement the meal you are eating. I don't think that we've purchased a vinaigrette at the store in the past 10 years. With how easy they are to make, there really is no need to.

In *On Food and Cooking*,[1] Harold McGee describes a slightly different vinaigrette-making method. He calls for adding the oil to a bowl first, followed by slow addition of the vinegar with vigorous whisking.

Is there a difference between these two recipes? Does the addition of oil to vinegar *versus* the addition of vinegar to oil really matter in the grand scheme of things?

Before we get to that, let's revisit this topic of hydrophobic and hydrophilic substances to understand the chemical reason for why they separate. Hydrophilic is the easier term to define. A substance that is hydrophilic means that it interacts with water in a way in which the two do not separate. Chemically speaking, that means the mixture is more stable, *i.e.* has a lower energy together, than either substance has on its own. Salt is a hydrophilic substance. When salt is added to water, it breaks down into sodium ions and chloride ions, which happily reside in the water. If sodium chloride-water mixtures were not stable, chunks of salt would just fall through the liquid and sit at the bottom of the container. (This is actually what happens when you add salt to olive oil.) There are some instances where sodium chloride is not stable in a watery liquid. When you add salt to a mixture that already contains a large amount of sodium chloride, the salt will not dissolve into the liquid. Water and sodium chloride still have a stabilizing reaction. But, that's not the right way to look at this. The total energy of sodium chloride and concentrated salt water (not regular water) is lower when the two are separate rather than when they've been confined. Because chemistry tends towards systems with lower total energy, chunks of sodium chloride will pass right through a saturated brine without dissolving. We have many examples of hydrophilic food substances from our everyday lives. Ethanol is hydrophilic. The vodka that you purchase from the store is a homogenous mixture; you don't see two layers in the vodka even though its 40% ethanol and 60% water. Sugars (like glucose, fructose, and sucrose) are also hydrophilic. Syrups

HYDROPHILIC

Hydrophilic chemicals have strong interactions with water

Weak Interactions
High Energy
Low Stability

Water and hydrophilic chemical, separate.
(High Energy)

Strong Interactions
Low Energy
High Stability

Water and hydrophilic chemical, mixed.
(Low Energy)

Hydrophilic interactions

The electrons tend to stay closer to the oxygen atom, giving it a slight negative charge

Positive things are attracted to the oxygen atom

The hydrogen atoms have a slight positive charge

Negative things are attracted to the hydrogen atoms

Chemicals that have strong interactions with water

Molecules with many –OH groups

Ionic compounds with charges

glucose   citric acid

$Na^{+-}Cl$ $NH_4^{+-}OH$

sodium chloride (table salt)

ammonium hydroxide

(whether its honey, maple syrup, corn syrup, or simple syrup) are products of the solubility of sugar in water.

There are chemical parts that cause these substances to have strong interactions with water. Water is a fairly simple molecule, a central oxygen atom surrounded by two hydrogen atoms. The hydrogen atoms stay close to the oxygen atom because of the way they share electrons with each other.

A discussion of electrons is critical to any real understanding of what is going on in chemistry. Electrons are the tiny, negatively-charged particles that zip around the inside of atoms and that you learned about as a teenager in your first (and for some their only) chemistry class. Electrons are the glue responsible for holding atoms together. Understanding how electrons move from one place to another in a chemical is necessary for understanding what that chemical does: is it sticky, is it toxic, is it nutritious, does it tend to turn into some other chemical, does it have strong interactions with water?

In water, oxygen shares two electrons with one of the hydrogen atoms and two electrons with the second hydrogen atom; our glue holding the bunch together. While the oxygen and hydrogen atoms do share the electrons, oxygen is greedier and the electrons end up spending most of their time by it. This stilted time-sharing results in a property called polarity; the oxygen is more negatively charged than the hydrogen atoms.

When I moved from California to Maryland, I had to pay taxes to both states. One of the questions that gets asked in the tax documentation is the date of my move. Because I moved in July, I was considered more of a resident of California and deemed only a part-time resident of the state of Maryland. One of the off-shoots of this is that I had to pay more money in taxes to California than Maryland. In the same sense, the bonding electrons in water are more of a resident of oxygen than of hydrogen. The electrons lend more of their negative charge to the oxygen and the outside world (*i.e.* other chemicals) view the oxygen in water as more negative than the hydrogen atoms (California was more Matt than Maryland).

Water tends to interact strongly with other polar chemicals. Other chemicals containing –OH groups tend to be polar and tend to have strong, stabilizing interactions with water. With 8 –OH groups, sucrose is an excellent example of this. Ionic compounds, things like sodium chloride, are extreme cases of polarity. In these chemicals, electrons aren't really shared as much

HYDROPHOBIC

Hydrophilic chemicals have strong interactions with water

Weak Interactions
High Energy
Low Stability

Water and hydrophobic chemical, mixed.
(High Energy)

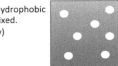

Strong Interactions
Low Energy
High Stability

Water and hydrophobic chemical, separate.
(Low Energy)

Hydrophobic interactions

The electrons in hydrophobic molecules are easily pushed around.

Molecules with this property have weak interactions with one another

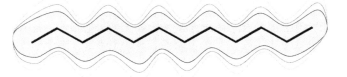

Chemicals that have weak interactions with water
(Molecules that have lots of carbon atoms and few OH groups or charges)

Flavor molecules

Oils

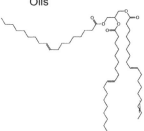

as horded. For sodium chloride, the "shared" electrons are only hanging out with the chloride and not between the two atoms. Because of this charge imbalance and the polarity it causes, ionic chemicals tend to be soluble in water.

A hydrophobic description is a completely different beast. Hydrophobic substances tend to not be hydrophilic. But, just because something isn't hydrophilic does not mean that it must be hydrophobic. There are chemists and materials scientists working on the development of materials that are omniphobic.[2] These types of materials are needed for coatings that do not attract water or oil. A playful demonstration of this is the coating that Joanna Aizenberg's lab made and placed in a ketchup bottle.[3] Being the skinflint that I am, I am excited to never again have to work to get every last gram ketchup out of a bottle before throwing it away.

It is true that hydrophobic chemicals are not polar; there is no electron gradient across the molecule. None of the atoms in the molecule hog electrons any more than any other. As such, the electrons in these molecules form a squishy cloud that hovers around all of the atoms. (This doesn't happen in a polar molecule because the electrons are more locked into position by the electron-greedy atoms.) When two molecules pass by each other, the squishiness of the electron cloud can stabilize the interaction between them. It turns out that these hydrophobic (non-polar) interactions are weaker than hydrophilic (polar) interactions.

Molecules that are hydrophobic are noted by their lack of –OH groups, or by the fact that they are not ionic compounds; things with charges are certainly not usually hydrophobic. Hydrophobic molecules do tend to primarily feature carbon and hydrogen atoms. Oils, which are hydrophobic, have many more carbon and hydrogen atoms than other kinds of elements. Glyceryl trioleate, one of the most abundant molecules in olive oil, is a molecule that is made up of 57 carbon atoms, 104 hydrogen atoms, and 6 oxygen atoms.

Of course, as with everything in this world, chemicals cannot always be neatly placed into bins of hydrophobic and hydrophilic, based on the fact that they contain –OH groups or based on the fact that they contain lots of carbon and hydrogen atoms. Sucrose (12 carbon atoms, 22 hydrogen atoms, and 11 oxygen atoms, with 8 sets of these H's and O's being tied up in –OH groups) is hydrophilic despite the fact that it contains 12 carbon atoms and 22 hydrogen atoms. Conversely, the presence of –OH groups does not mean that a molecule will be hydrophilic. Methanol (CH_3–OH), ethanol

(C_2H_5–OH), and propanol (C_3H_7–OH), are all hydrophilic as measured by their solubility in water (the mass of a substance that will dissolve into water without separating). Butanol (C_4H_9–OH) with a solubility at 8 grams per 100 mL of water, pentanol (C_5H_{11}–OH) with a solubility of 2 grams per 100 mL of water, and hexanol (C_6H_{13}–OH) with a solubility of 0.7 grams per 100 mL of water, would all be considered hydrophobic.[4] Seemingly what's important here is the number of –OH groups per carbon, with a ratio of 1 –OH group for every 3 carbon atoms being a cutoff between what is hydrophilic and what is not. Of course this isn't technically true for every case, but it can be a guidepost for quick decisions when trying to figure out if a molecule will be hydrophilic or hydrophobic.

But, there is a real and measurable attraction between hydrophilic chemicals, just like there is a real and measurable attraction between hydrophobic chemicals. The observed repulsion between hydrophobic and hydrophilic materials is really just the result of the fact that they are more stable (have a lower energy) on their own.

In chemistry circles, we often describe this by saying, "Like attracts like." Of course, this refers to hydrophobic attracting hydrophobic and hydrophilic attracting hydrophilic. Another way to look at it is that molecules that look like one another tend to be attracted. The polar group on the water (–OH) is the same as the polar group on sucrose (–OH). They look like each other and are attracted. Oils are attracted to one another because they look alike; they have the same makeup.

Imagine a cup of water that has a single blob of oil floating through it. Mostly, the oil is fine enough floating around in water, it is repelled in one spot as much as it is repelled in the next. By that token, it could go anywhere it wanted to within the cup of water. Eventually though, that blob rises to the top (dense things like water sink and less dense things like oil will rise) and will flatten out. Now imagine two droplets of oil in the cup of water, each at opposite ends of the cup. Again, those two droplets are happy enough swimming around. Eventually, one of two things will happen. They will both, of their own accord, rise to the top, find each other, and flatten out. Or, they will find each other and then rise to the top and flatten out.

The point of this description is to point out that they think they are OK floating around in water until they realize that the other one exists. Each droplet has to randomly move around the cup

OIL & WATER

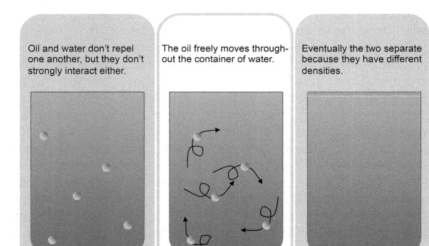

until they randomly bump into one another. There is no long-term attraction that pulls them close together. Oil doesn't have a magnetic feel that searches out other oil. It is movement that eventually bears out the fate of all oil in water. It will separate eventually.

McGee describes the proper way to make a vinaigrette in *On Food and Cooking*. Starting off with some olive oil in a bowl, slowly add vinegar. At first, the vinegar should be added drop-by-drop with plenty of whisking. Eventually, you will notice that the mixture has become thicker (it is more viscous and doesn't slosh around as much while whisking) than the oil on its own. At this point, you can add the vinegar a little faster. Keep adding the vinegar until you get to the balance of oil and acidity that you want. While making your vinaigrette, you need to be wary of its consistency. If it is to thick, you need to add more oil or else it will break, the oil and acid separating. If it is too thin, you need to add more vinegar or else it will break. Knowing what the proper thickness and proper ratio is a matter of trial and error and experience.

In this example, vinegar gets added and is knocked into little droplets that get spread throughout the oil. The vinegar is the dispersed phase. The oil is the continuous phase. And the mixture of the two is called an emulsion, which is just a generic term for a stable mixture of a hydrophobic substance with a hydrophilic substance.

Vinaigrette

There are lots of different kinds of emulsions in the culinary world. Coffee is an emulsion (the essential oils and aromas of the coffee bean are suspended in water). Butter is an emulsion (water droplets and proteins are suspended in milkfat). Milk is an emulsion (fat particles are suspended in water). Egg yolks are an emulsion (fats and cholesterol suspended in water). Mayonnaise is an emulsion (oil suspended in water, supported by molecules from an egg). Vinaigrettes (the on-line and recipe book variety) are emulsions (oil suspended in acid). And, vinaigrettes (the Harold McGee variety) are emulsions (acid suspended in oil).

In each of these cases, a dispersed phase is suspended in a continuous phase. The dispersed phase takes on the shape of little droplets, spread throughout the liquid. The continuous phase encompasses all of these droplets. In the best emulsions, the continuous phase is only visible as a narrow band around each individual droplet of the dispersed phase.

As you can see from this description, the dispersed phase can be either hydrophobic or hydrophilic. There is no hard and steady rule that it has to be one way or another. But, the rules for making a vinaigrette are set by which material is the dispersed phase and which material is the continuous phase.

When making an emulsion, it is important that all of the ingredients are at the same temperature. The first step is to add the continuous phase to the bowl. This is also a good time to steep any extra aromatics you might want to flavor your vinaigrette. These can be strained off before beginning the emulsification. The next step is to start adding the dispersed phase, drop-by-drop. You really need to add slowly at this point with plenty of whisking; all of the droplets of oil need to remain small and stay spread out through the liquid. It is imperative that they not meet up to form bigger droplets. In this early stage, the viscosity of the vinaigrette is too thin, all of the molecules in the liquid are able to slosh about quickly. The slow addition really makes sure that no big dispersed-phase blobs form and break your emulsion.

As it starts to develop, the emulsion becomes noticeably thicker. This happens because the oil droplets start packing in closer and closer together. Eventually the pattern that they make looks something like a honeycomb. In this analogy, the open spaces of the honeycomb comb are the dispersed phase, and the solid portion corresponds to the continuous phase.

Let's think of a single blob of dispersed phase again, and what it experiences as the emulsion thickens. Just like when there is a single droplet of dispersed phase in the container, each individual blob wants to randomly move around through the continuous phase. But, in our emulsified state, when the blob tries to move, the continuous phase tries to move out of its way. The only problem with that is that any direction the continuous phase tries to move in, there is another blob of dispersed phase standing in its way. As the dispersed phase packs in tighter and tighter, it creates its own straight jacket, its own personal jail. Any molecular movement would lead to increased interaction between hydrophobic and hydrophilic materials.

In chemistry, the goings on at the molecular level always work themselves up to create effects that we can detect with our own senses. Molecular motion is unambiguously related to whether a substance is a solid, a liquid, or a gas. In a frozen solid, the individual chemical units are frozen in space, unable to move up, down, left, right, backwards, or forwards. It is solid because there is no motion. For liquids, there is some movement. The ocean can form waves because the molecules in the sea are able to move. Gases, then, are made up of chemicals that buzz around really quickly and don't hold any particular shape.

For our vinaigrette, the hydrophobic-hydrophilic trap that it finds itself in results in molecules that are no longer able to move as easily as they once did. The molecules slow down. The vinaigrette becomes more solid. An expert chef can bring a good vinaigrette to a weak jelly-like state.

Eventhough the vinaigrette seems solid, it is fragile. The system energetics dictate that the emulsion will eventually separate. The hydrophobic parts will eventually all pool together, as will the hydrophilic parts. When it's made however, the continuous and dispersed phases are separated by a kinetic barrier. That is, room temperature motions are not enough to quickly take the emulsion down. But a singular jolt that is strong enough and sharp enough, can jostle the molecules just enough so that the two phases start to separate.

The other thing that you need to be really careful of when making an emulsion is that it does not become too firm. Again, firmness indicates that the blobs start packing more and more densely. Try to add any more, and the droplets will run out of room, start to coalesce around one another, and the emulsion will separate. I have personally never reached this stage when making a vinaigrette, but it is possible. Should you ever encounter this problem, a drop or two of continuous phase will bring balance back to the emulsion.

These rules apply for other emulsions as well. One of the most beautiful emulsions that I have ever seen is aioli. Real aioli. Not the stuff we Americans think about when we think of aioli: mayonnaise with added garlic. That is absolutely not what real aioli is.

Aioli has 4 ingredients: garlic, salt, olive oil, and a sparing amount of water, if necessary. I have always been too intimidated to make aioli on my own. When done right, it can easily take 30 minutes of constant motion. It takes dedication and experience to make it, which is one of the reasons why I have never tried.

Harvard University has an immensely popular science of cooking class that is run by faculty members Michael Brenner, David Weitz, and Pia Sörensen. I am incredibly envious of this class.[5] They bring in world-famous chefs to lecture to their students and to give public seminars. Many of these have been recorded and are disseminated through services like iTunes U. Being an avid fan of cooking science, I have actively devoured these videos. Of all of them, my favorite is the lecture given by Nandu Jubany, a Catalan chef and restaurateur who runs the renowned Can Jubany restaurant, among others.[6] This particular lecture is all about emulsions, although they do go over other types of thickened sauces as well. Beyond impressive, to me, is the way he makes aioli. He crushes 10 garlic cloves and a heaping pinch of salt into a paste with a mortar and pestle, after which he starts slowly adding oil. He adds the oil by the drop in conjunction with constant motion from the pestle. Following along with the rules of making an emulsion, he adds the oil quicker as the emulsion starts to set up. All in all, he adds 1 liter of oil in making the aioli. I never tire of watching this. The patience and understanding that are fully evident as he starts making the aioli and the solidity of the final product. Even though I understand the science, seeing a full liter of oil set up like that into a semi-solid is nothing short of amazing.

And the stability of these sauces hangs on a knife's edge. Bumping the container the wrong way, changing the temperature, looking at it cross-eyed, neglecting to howl at the moon, and forgetting to rub your rabbit's foot three times for luck, are all ways to quickly break your emulsion and see the dispersed phase fall out of suspension.

All emulsions will eventually share this fate. Give them enough time, and they will break. The beautiful diffraction patterns we see in rain slicks are also plainly visible in a cup of coffee that has sat out too long. Oils, sheltered away in water, will eventually find each other. Stabilizing emulsions is so inherent in many of our recipes that we take it for granted.

One group of people who certainly do not take emulsion stability for granted are food scientists. The food industry also cares a great deal about this. I would guess that, of all of their activities, more than half of all food scientists perform research that is directly concerned with emulsion stability, which is directly related to how long that food will remain fresh. Most of the foods that we eat have a balance between hydrophobic molecules and hydrophilic molecules. The hydrophilic molecules tend to be the delivery vehicles. As our planet is mostly water, it should come as no surprise that our foods also contain a great deal of water. But water on its own holds no interest to us. All of the things that make food interesting and attractive tend to involve hydrophobic molecules. Flavors, aromas, textures, and colors all tend to come from hydrophobic molecules.

How do we maintain that balance? How do we ensure that our food products stay fresh at the grocery store? How do we keep food waste to an absolute minimum? For many foods, there is chemical intervention that is necessary to optimize shelf-stability.

Many claim that we should only ever eat fresh foods, and that all processing is unhealthy for us. I strongly disagree with that sentiment. As we push towards a population of 8 billion, our ability to feed all people will hinge on utilizing every bit of food that we grow. Because some of that food necessarily has to move to places far-off, we will need processing methods that keep the food free from pathogens, while maintaining freshness, for as long as possible.

Since it is the least complicated, though it is certainly not the most crucial or important, I will focus on the example of sports beverages. As we all know, sports beverages are filled with the electrolytes our body needs after strenuous exercise. (This statement, of course, is mostly hyperbole. For in depth satire on our love of electrolytes, please see the movie *Idiocracy*.[7]) Anyway, most of the chemicals in a sports drink are just fine hanging out in water; all of those electrolytes come from ionic compounds, which have charges and are hydrophilic. All of the colors and most of the flavors, however, are anything but hydrophilic. They will fall out of water the first chance that they get. Traditionally, manufacturers of sports drinks used brominated vegetable oils to stabilize the colors and flavors. Brominated vegetable oils are just vegetable oils that have been, well, brominated. In 2013, an on-line petition sought to have brominated vegetable oil removed from all formulations.[8] The petition was started out due to potential health concerns over brominated vegetable oil. And, while it is effective at it's role in the drinks and did not pose any real health risk, the food industry needs to be hyper-aware and sympathetic to their customers.

Brominated vegetable oil wasn't the first food ingredient to be questioned by the public and it certainly won't be the last. But the fight against it is representative of the way that many people speak out against chemicals in their food. As a chemist, my first reaction to these campaigns is usually pretty haughty; I associate with other scientists and I am quick to both understand and side with the research of the scientists working in the food industry. My initial instinct is to want to feel disdain for the people who question that work. But, these reactions and feelings on my part, are not only foolish and counterproductive, these feelings are wrong. The people who question chemical additives have every right to do so. When looking at the molecules in our food, I can pretty quickly figure out misinformation and misunderstandings from fact. But, I can only do that because I actively and knowingly work with chemicals everyday. I am accustomed to thinking about how they move and function and interact with our bodies. There is not an inconsequential number of people who are only willing to use "chemical free" products in their lives. As a chemist, it's easy for me to say, "Everything is a chemical," and look down my nose. But, when was the last time that most people actively thought about the chemicals that they handle on an everyday basis? When they took a chemistry class? It is an utter failure on the part of chemists to engage people with chemistry, that we have now reached a point where marketers can be successful selling "chemical free" products. People only want to have a little control over their lives, and as a whole, we chemists have actively built up a pretty high barrier between the way that we engage with chemicals and the way that non-chemists engage with chemicals. This is a shame, and I hope that as a profession, chemists will do a better job, of getting people not just to accept our research, but also in finding ways to actively engage the public to be part of our work. The food industry and the chemical industry, more broadly, would do well to facilitate that engagement.

The removal of brominated vegetable oil from sports drinks necessitated a new addition. If this molecule could keep the flavors and colors from separating from water, brominated vegetable oil would have to be replaced. (My favorite color and flavor for these drinks is purple. And, yes, in this case, I consider purple to be the most accurate description for that flavor.) For some companies, the chemical of choice became sucrose acetate isobutyrate.

What is it that these chemicals do? And, why are they necessary?

Both of these molecules keep the flavor and color compounds from interacting with one another. If all of the chemicals that are responsible for color started aggregating, these drinks would have little pockets of concentrated purple (or orange or red or blue) floating around in the middle of a clear liquid. Actually, it is likely that these aggregates would rise to the top of the liquid or fall to the bottom, depending on their density. Both brominated vegetable oil and sucrose acetate isobutyrate, also help to ensure that any aggregates that might form, stay well distributed throughout the entire container and don't find their way to the top or bottom.

Generically, these molecules are called emulsifiers; an emulsifier being anything that helps to stabilize an emulsion. Within this broad classification, there are emulsifiers that work by exploiting all sorts of different chemical properties. The brominated vegetable oil and sucrose acetate isobutyrate focus solely on interactions with hydrophobic molecules. Another type of emulsifier includes hydrocolloids, which I have covered more broadly in the chapter on jelly. Briefly, hydrocolloids work by thickening an emulsion, slowing down molecular motion, and, more specifically, by creating strong interactions with any water in the mixture. A final class that might be the most effective, are amphiphilic emulsifiers.

Emulsifiers & Emulsifications

An emulsion will always separate. But, you can keep them stable for a long time by playing a few tricks.

Packing tiny droplets of one phase within another will prevent both the oil and the water from moving and, ultimately, separating.

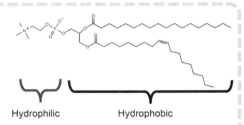

Hydrophilic   Hydrophobic

Amphiphilic emulsifiers, like lectithin, coat the droplets in an emulsification, preventing water and oil from interacting.

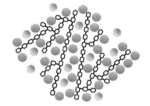

Hydrocolloid based emulsifiers, like xanthan gum, slow down molecular motion, and delay the point when the water and oil are able to separate.

Amphiphilic is just a fancy chemical word that means it is a longish molecule (bigger than typical flavor and aroma compounds) that has parts that are distinctly hydrophobic and parts that are distinctly hydrophilic. That description is a bit of a mouthful, which is why we usually just say amphiphilic.

To get an idea of how amphiphilic emulsifiers work, let's talk about soap. Any dish or laundry detergent will contain surfactants, which maybe you have heard of, and which are also amphiphilic emulsifiers. Imagine throwing a grease-stained shirt into the laundry, something that all of us have done before. We instinctively know we would need to add soap to clean that shirt. Water alone will not work. The reason for that is the absence of chemical attraction between the water and that spot of grease. The water is hydrophilic, obviously. And, the grease is hydrophobic. The grease would much rather cling to your shirt than it would slip into the water. This is where soap, and its chemical identity as an amphiphilic emulsifier, come into play. When you put soap into your washing machine, the individual molecules of surfactant get mixed through the water. Some of them will find their way onto that grease stain. Once there, the hydrophobic ends of the molecule will start interacting with the grease. Eventually enough surfactant will find its way to that stain so that the whole thing gets covered in soap molecules. The soap molecules can even displace the grease from the fabric as its hydrophobic ends replace the interactions between the grease and your clothes. You will end up with a blob of grease that has the hydrophobic sides of the amphiphilic emulsifiers attached to it. Now, because these emulsifiers have two ends, we also need to think about what its hydrophilic ends are doing. The hydrophilic parts are, for all intents and purposes, eternally attached to the hydrophobic parts. They will never be able to escape each other. So, if the hydrophobic end is facing one direction, the hydrophilic end must face the other. For our soap molecules, this leads to a happy circumstance. While the hydrophobic parts point inwards toward the grease, the hydrophilic parts point outwards toward the awaiting water in the washing machine. The water, at this point, has no sense that there is a big grease ball in the middle. The water can only tell that there is a blob covered with hydrophilic molecules. The water is perfectly content to interact with those hydrophilic parts, the grease gets suspended in the basin of the washing machine, and the water carries the emulsifier-supported grease down the drain.

Chemical bonding and interactions are often difficult to understand, even for a chemist. In fact, bonding and its definitions make up the playground for chemical philosophers. "What is a bond?" "What does it mean for two molecules to be bonded?" There are no right answers to these questions. A chemical bond is a fictitious construct; made up by chemists to help us best describe what we see. What is so different between the interaction between two adjacent carbon atoms in the same surfactant molecule and the interaction between one surfactant molecule and a grease molecule? Is it the permanence of the interaction? The two carbon atoms may be torn apart in a more violent chemical reaction while the interaction between two separate molecules may be more fleeting. Does the constancy of an interaction make it a bond? What about two molecules that are held together like two links in a chain? These two pieces aren't actually attached to each other, but it would take a lot of energy to tear them apart. Does that mean that, even though they don't actually touch each other, they are bonded? For a chemist, these are fun to think about. But for the purposes of this book, we view a molecule, like an amphiphilic emulsifier, to have a permanence that the interaction between two separate emulsifiers just can't have. Of course the bonds between atoms in a molecule can be broken. We see this in the high temperature processes that occur in the Maillard reaction, which generates new flavor molecules from proteins and sugars. But, for the mild temperatures at which emulsions are formed, a single molecule, like an amphiphilic emulsifier, will not break apart.

Amphiphilic emulsifiers work in our foods the same way that they work in your laundry machine. The hydrophobic ends direct themselves toward hydrophobic things, and the hydrophilic ends direct themselves towards hydrophilic things. Having a peek at our vinaigrette, the emulsifiers will just line themselves up along the contours of the honeycomb pattern. If oil is the dispersed phase, the emulsifiers will stick their hydrophobic parts into the middle of the droplet while the hydrophilic parts poke their heads out towards the vinegar. In an emulsification that is stabilized like this, the continuous phase has no knowledge that the dispersed phase is even there. All that the hydrophilic continuous phase interacts with is the hydrophilic ends of the emulsifiers, and it is just pleased as punch to do so. The amphiphilic emulsifiers, then, make a sort of chemical bridge that separates but connects hydrophobic to hydrophilic.

This chemical bridging has dramatic results in the formation of mayonnaise. In recipes for mayo, the amphiphilic emulsifiers from egg yolks are used to stabilize the water and oil mixture. Egg yolks contain two kinds of emulsifiers. The first includes all of the proteins. As there are some amino acids, the pieces that make up all proteins, that are considered hydrophobic, and some that are considered hydrophilic, all proteins are naturally amphiphilic. Egg yolks also contain a molecule called lecithin, which is the prototypical-emulsifying agent.

   A typical mayonnaise recipe is as follows. Use an immersion blender and a cup in which the blender fits snugly. To the cup, add 1 egg yolk, 15 mL of water, 15 mL of lemon juice (I substitute 1/4 teaspoon of citric acid and 15 mL of water for a cleaner taste), a pinch of salt, some mustard or cayenne pepper or other flavoring for taste, and 1 cup of canola oil. Place the blender into the cup and start to slowly pulse the mixture. After several good pulses, when you can see that the mayonnaise has started to firm up at the bottom of the cup, you can start to slowly lift the blender through the entire liquid in the cup. All in all, this process should take about two minutes. You can make mayonnaise in a bowl with a whisk as well. Just be sure to add the oil slowly at first. Either way, you need to follow the generic rules for making emulsions. What most amazes me about mayonnaise is that so much liquid oil can be made to firm-up as a semisolid with just a bit of egg and water; with about 30 billion separate oil droplets constituting 15 mL of mayonnaise. Harold McGee claims that a single egg yolk can emulsify many cups of oil. Just incredible.

BACK TO THE BASICS

We still haven't answered the question of how to best make a vinaigrette. What is the best choice for the dispersed phase and the continuous phase? In McGee's recipe, one of the reasons why he uses olive oil as his continuous phase is that it is more viscous and that viscosity will cause the emulsion to keep longer.

   Kenji Lopez-Alt has also done some experimentation with vinaigrettes.[9] He found, adding just oil or vinegar to leaves of lettuce, that the oil seeps into the lettuce, causing it to look wilted. The vinegar, on the other hand, rolls right off. This has to do with the waxy film, a product of nature, that coats every leaf of lettuce. The oil, being hydrophobic, can interact with, penetrate,

Vinaigrette 153

and loosen the wax, which is also hydrophobic. The vinegar just rolled right off. Kenji did another experiment. He made 2 vinaigrettes with both using a blender. (The pictures that he shows indicate that the resulting emulsion is a vinegar in oil emulsion. I would guess that the higher viscosity of the oil preferentially caused it to become the continuous phase; the vinegar, with its lower viscosity, being more easily dispersed.) The first vinaigrette contained just oil and red wine vinegar. The second vinaigrette contained oil and red wine vinegar and mustard. Mustard contains a mixture of several emulsifiers, some hydrocolloids and some proteins, which together form mucilage. The mustard stabilizes the emulsion. It also tempers how the oil interacts with the lettuce leaves. He found that the oil in the emulsion held onto the leaves just tightly enough so that it could cling to the surface without interacting so strongly, so that the oil could penetrate the waxy parts and wilt the lettuce.

So what's the right way to make a vinaigrette? I have always made mine using vinegar as the continuous phase. That's the way that most chefs make it. That's the way that Nandu Jubany makes it in his video for Harvard's Science of Cooking class. If you make it thick enough, it will nicely coat your salad. But making a thick vinaigrette is not for the faint of heart or inexperienced. If you want a vinaigrette that is easy to make and clings to your lettuce in just the right way, it is probably best to have oil be the continuous phase in a stabilized emulsion.

I may just have to change how I've always made vinaigrettes.

REFERENCES

1. H. McGee, *On Food and Cooking*, Scribner, New York City, 2004.
2. Z. L. Chu and S. Seeger, Superamphiphobic Surfaces, *Chem. Soc. Rev.*, 2014, 43(8), 2784–2798.
3. J. Aizenberg, *Extreme Biomimietics (TEDxBigApple)*, [video file]. 2012 March 2 [cited 2016 July 5], available from: https://www.youtube.com/watch?v=nVOzkO-ccuc.
4. J. Altig, *Solubility of Alcohols*, [document on the Internet]. NMT.edu [cited 2016 July 5], available from http://infohost.nmt.edu/~jaltig/SolubilityAlcohols.pdf.
5. Harvard University, *Science & Cooking*, [document on the Internet]. Harvard.edu [cited 2016 July 5], available from: https://www.seas.harvard.edu/cooking/cooking_archive.

6. N. Jubany, *Emulsions: Concept of Stabilizing Oil and Water (Harvard University)*, [video file]. 2010 November 12 [cited 2016 July 5], available from: https://www.youtube.com/watch?v=fU4sbFgTicg.
7. *Idiocracy*, Hollywood, Ternion, 2006, Film, director Mike Judge.
8. R. Burks and B. Valsler, *Brominated Vegetable Oil*, [podcast on the Internet]. Royal Society of Chemistry; 2016 March 2 [cited 2016 July 5], available from: http://www.rsc.org/chemistryworld/2016/03/brominated-vegetable-oil-podcast.
9. K. Lopez-Alt, *What's the Point of a Vinaigrette?*, [document on the Internet]. Serious Eats. 2010 April 16 [cited 2016 July 5], available from: http://www.seriouseats.com/2010/04/salad-dressings-vinaigrettes-the-food-lab.html.

Dinner

CHAPTER 9

Pizza

What is the secret to good pizza?
   That's easy.
   Water.
   The secret to making good pizza is water.
   I'm not talking about water in the way that some coffee people and bagel people talk about water. In their telling, water has to have exactly the right mineral content (general hardness as a measure of the amount of magnesium and calcium) and the exact right alkalinity (pH as determined by the concentration of bicarbonate in the water). There are some who believe that good bagels can only be made in New York City because of the ideal (for bagel-making) mineral content of its water. Christopher Hendon (a chemist) and Maxwell Colonna-Dashwood (the owner of a specialty coffee shop) have actually defined the ideal hardness for brewing coffee and have written about ways for people to alter their own water to make it ideal for making coffee at home.[1]
   No. The type of water and its purity are not on my list of things that I think about when I'm making pizza.
   Over time, I have developed several philosophies that guide my pizza making. But the most important philosophy that I have is that the water in my pizza must find the proper balance.

The addition of water is necessary for every ingredient. The tomatoes (for the sauce) rely on water to grow and ripen. Brining mozzarella in salt water imbues with flavor and just the right texture. And dough, no matter what kind you make, primarily contains flour and water.

Balance comes when the baked pizza has the taste and texture that you want. Crispiness or chewiness of the crust. How melted the cheese becomes. The thickness of the sauce. Getting what you want out of your pizza requires that you understand the amount of water in your ingredients and how you need to bake your pizza, which removes water from the ingredients, to achieve the textural qualities that you desire.

WATER IN. WATER OUT. BALANCE

My own quest for finding balance in pizza making (much like finding balance with myself) has run over the course of my life, and I certainly expect it to continue as my tastes and preferences change.

I think that the first pizza I ever made was probably an English muffin pizza. I expect that this is the same for most of you. For any reader who is not familiar with this culinary delight, a brief description is in order. Half of an English muffin is topped with pre-made pizza sauce (typically purchased from the store). Shredded mozzarella cheese is sprinkled onto this makeshift pizza, which is then baked in the oven until the cheese melts and the muffin gets toasty. It is, in a word, incredible.

What kid doesn't love making their own pizzas? My kids certainly do.

Rare is the day when all of my kids are happy with what gets served at the dinner table. My wife and I have to put up with a lot of whining and complaining. There's always one kid (and its always a different one) who feels the need to gripe about what they're supposed to eat. Sometimes, it gets to be unbearable. I can recall a time when I was travelling for a conference. I called home to chat and to see how Erika and kids were doing. (I always half-dread making these calls because if the kids are being difficult, there's nothing I can do to help.) On this particular occasion, Erika was completely fed up with the kids, and just wanted a moment's peace for herself. She told me that she couldn't take it anymore. No one was happy about anything. Later I found out that she went to the store, bought some English muffins, cheap pizza sauce, and cheep mozzarella cheese. The kids were ecstatic and remained completely reasonable for the rest of the day.

Even the simplest and most humble of pizzas has the power to enchant.

There are many regional variations on pizza throughout the United States. This variation is highlighted by the stark difference between New York style pizza and Chicago style pizza. New York style pizza is noted for its foldable crust, which has a slight crispness on its edges. The pizza is lightly sauced and often sold by the slice. Chicago style pizza, on the other hand, is not even considered a pizza by many. The dough for this pizza is formed and shaped within a pan; the depth is necessary to hold the bounty of cheese and sauce and any other ingredients that might be able to fit within its walls. It is a meal that is not for the calorie conscious.

Of course, all pizzas are held to the standard *pizza napoletana*, which has a strict set of requirements for how it must be made. The dough is made with only four ingredients: soft Italian flour (type 0 or 00), water, salt, and yeast. The dough is hand stretched to 3 mm, coated in sauce made from San Marzano tomatoes, and topped with mozzarella (*mozzarella di bufala Campana*). The pizza is quickly baked in a stone oven, heated by a wood fire [to near 500 °C (930 °F)]. Because of the adherence to these specifications, there is even a group, *Associazione Verace Pizza Napoletana*, who will certify a *pizza napoletana* as being *Denominazione di Origine Controllata* (controlled designation of origin).

But a brilliant idea, such as pizza, cannot be contained within a single definition, no matter how delicious that definition is. Even in the US, the number of types of pizza is a varied bounty of goodness. New Haven pizza (a close relative of New York pizza) has a slightly crisper crust, with a more blackened and crunchier exterior, and a minimal use of cheese. California pizza is a single-serving sized pizza that tends to feature fresh and local ingredients, which aren't necessarily typical of most pizzas. St. Louis pizza has a cracker-crispy crust, which is made from dough that has no yeast. The pizza is round, but is cut into square-shaped pieces (at least most of the pieces are square).

The pizza where I grew up, near Dayton, Ohio, is similar to St. Louis pizza in that its crust doesn't rise like Neapolitan-style pizzas and is cut into squares. The key difference is that the crust isn't cracker-crisp and can have some baking powder or yeast to produce some small amount of rising. There are several local chains that I grew up with: Cassano's, Marion's, and Donato's, to name but a few. And these pizzas share a few traits that stand out to me. The first are those square slices. I know that when most people think of pizza, they immediately think of a triangular slice of pizza. There is something special, however, about the square slice. The small size is tantalizing because you always think that you could just have another, and another, and another. (Some might think this a bad quality, but I don't think so.) For the pizza, as a whole, the toppings go right to the edge of the crust. Again, some may think that this is heresy, that the crust plays too important a role to not be prominently featured. However, when the crust is seasoned properly (some salt and a little yeast or some other herbs and spices to add flavor), the crust and sauce and toppings are in ideal proportion with one another. Every bite is perfect.

Growing up, pizza night was always special. I imagine that this is the case for most households. Mom and Dad loved it because they didn't have to make anything and everybody was happy about the results. As kids, we loved it too, because we just liked pizza so much. My little brother Tom was the picky eater who needed his food prepared to order (the right food, cooked the right way, served the right way). He wasn't demanding. I think he ate peanut butter and jelly for lunch every day between the ages of 6 and 18. But, he would bend over backwards to get pizza.

Whenever our parents would deem to ask us what we wanted for dinner, pizza was always Tom's answer. On Christmas Eve, my family always made a bunch of "small bite" foods. Everyone gets to pick what special food they would like. I would pick something like shrimp or stuffed mushrooms. My parents stopped asking Tom what he wanted by the time he turned 10. It never mattered where the pizza came from either. Most likely it would come from Cassano's (the local Cassano's in my home town changed its name to Beppo Uno somewhat recently) or Little Caesar's (a Detroit-based chain who would sell their pizzas two at a time). But, we would get frozen pizza too (I remember Red Baron's, Tony's, DiGiorno, and Freschetta).

Left to his own devices, Tom ate his fare share of pizza while at college too. I asked him about this recently. He told me that he probably ate pizza 5 out of every 7 days in the week. Most of that was frozen pizza (Tony's or Red Barron's, whatever he had a coupon for). But, every Tuesday, he and his roommates would order from Domino's because of the deals they had that night.

It's funny how we remember these things, seemingly inconsequential, that actually define who we are. My college roommates and I would always get wings on Tuesday nights because of they were cheaper then. The wings really weren't memorable in the culinary sense. But they were a shared meal, an expected indulging that we always looked forward to.

I asked Tom what his favorite pizza is. Without hesitation, he told me Giordano's, from the Chicago stuffed-crust pizza chain. Having lived in Chicago, I have seen Chicago people defend their pies from people who would denigrate them as "not actual pizza." Tom never lived in Chicago, though he did visit me there every now and then. So, I was a little surprised when he told me this. It's not often you hear a non-Chicagoan saying that a Chicago-style pizza is their absolute favorite. I asked him why. When he answered, I could hear his incredulity over the phone. "Seriously, Matt. You're an idiot. It's really simple," his words and tone jokingly implied. "There is more pizza in every bite than any other style of pizza out there" I've got no argument for that!

I was no stranger to pizza as sustenance in college either. Domino's pizza always had the best deals for students at the University of Dayton. So, that's where my roommates and I would order. Every now and then I would be treated with a trip to Marion's. But, what I relied on most was Jack's pizza. Jack's is a brand of frozen pizzas. Not only was it one of the cheapest options (at the time, Jack's cost $2.50 for a pizza while Red Barron's or Tony's cost closer to $5 per pizza), it was the best. I loved Jack's so much that I would still buy it after I was married, on nights when my wife was busy or eating with friends or when I would come home late from the lab. It has been a while since I've had a Jack's pizza, although I recently found a retailer, only several kilometers away, who sells it.

As with everything I cook, my preferences and tastes for pizza have been informed by the places I have passed through (and the types of pizza sold there). After Dayton, I attended Northwestern, which is in a suburb just to the north of Chicago. Everyone in Chicago has an opinion on which Chicago-style pizza chain they liked best. The chain I preferred has always been (and will always be) Giordano's. There is something about their buttery crust that Lou Malnati's and Geno's East just can't compete with. My favorite pie, at a shop close to where Erika and I lived, was BoJono's. After Chicago, we moved so that I could take a position as a postdoctoral researcher at the California Institute of Technology in Pasadena, California. The cafeteria at Caltech sold some really nice personal-sized pizzas whose crust was made from un-yeasted dough. Whenever we would go to the beach, Erika and I usually chose to go to Hermosa Beach because that's where Paisano's pizza is. Paisano's does up a really nice New York style pizza that you can buy by the slice or as a whole pie. We would usually buy a whole pie and bring home the leftovers. A favorite of people at Caltech was a place called Zello's. Zello's makes a deep dish corn meal crust-based pizza. Their version with caramelized onions, corn, and Gorgonzola is to-die-for-good. While in the LA area, my wife and I even got to go to Pizzeria Mozza, which is co-owned by celebrity chef Mario Batali. This is still the home of the best mozzarella and burrata I have ever had. On our move from LA to Washington DC, where we currently live, I did a cross-country drive with my brother Tom. He and I stopped in Phoenix to stay with our sister, Johanna, and her husband, Kyle, and were lucky enough to eat at Pizzeria Bianco, which does Neapolitan-style pizzas and is often ranked as one of the best pizzerias in the country. I still count the salami pizza that I had there as the best that I've ever had. The quality of the crust and sauce and cheese, coupled with the spice and saltiness of the salami, which had started to fry in its own fat in the hot oven, all came together in a really memorable way for me.

Now that I live in DC, my family and I have two favorite places. Sadly (for our taste buds), both of these are closer to work than our home. (Perhaps this is a good thing for my bank account, though.) The first is Pete's New Haven Style Apizza. They have a really great pie with fried eggplant and spinach and onions. The other is 2 Amy's pizza, which does a bustling business with their *pizza napoletana*.

Erika and I really don't go out for pizza as much as we used to anymore. This might be because we don't have a go-to place that's close to home. (This would be different if we lived closer to 2 Amy's and Pete's.) It might also be because I've figured out how to make a decent enough pizza on my own.

CRUST

A great pizza starts with a great foundation. My personal preferences really define what I want in a pizza. With my own pizza, I try as hard as I can to approximate something close to a Neapolitan pizza: thin and pliant (not soggy or crunchy) undercarriage (bottom of the crust) with some spotting and that displays a good amount of oven spring (it fluffs up) around the edges.

In contrast, a New York style pizza cooks longer, doesn't have an oven spring that is as pronounced, and has a crust that is a little heartier (read: thicker). Uncooked dough for a Neapolitan pizza also spreads more quickly and more easily than dough for a New York style pizza. My friend John, a high school chemistry teacher, makes a really nice New York style pizza using a dough recipe from America's Test Kitchen.[2] Kenji Lopez-Alt has also written very useful pieces about making New York style pizzas at home.[3]

The first time I tried to make a pizza was unmemorable. Well, it was unmemorable to me. This could be my selective memory kicking in because the pizza I made wasn't so good. Or it could be because the pizza I made was just ho-hum. Either way, I do remember that I used a pre-boxed flour mix (probably Jiffy brand) to make the dough. This probably happened when I was a teenager, living with my parents. I don't' think I tried any sort of pizza making after that until I was in my mid 20's. I am guessing that this attempt used pre-made dough from Trader Joes, a national grocery chain. At the time, I had heard from other people who used and enjoyed this dough. The pizza that I made was certainly better than what I had put together with the boxed flour mix. Of course, this could have been because I was older and more experienced. But, Erika and I stuck with this dough for quite a while when making pizza at home. This is still my father's go-to when he makes pizza for Mom and their friends and family. It is definitely serviceable.

I probably would have continued to use this pre-made dough had I not started baking my own bread. Like many, I got my start making bread inspired and emboldened after reading Mark Bittman's description of Jim Lahey's no-knead bread recipe in *The New York Times*.[4] A brief description: combine dry ingredients (flour, salt, and yeast), add water and mix until no dry flour remains, allow to sit out overnight while covered, form into desired shape, allow to rise for two hours, bake in a pre-heated cast iron pot with a lid.

This recipe sparked a real resurgence of bread making at home. The attraction, to many, was the no-knead descriptor. Kneading always seems like a tedious chore and, for the untrained, there is a real uncertainty to knowing when you are finished. There is also some art to kneading. To me, though, there are two genius parts to this recipe. The first involves baking your bread inside of a Dutch oven. Professional bakers have really tight control over both the temperature and humidity of their ovens. It turns out that quality bread requires a humid environment. When you bake your bread inside of a heated Dutch oven, the lid traps the water escaping from the bread as it is warmed, approximating the inside of a baker's oven. The second genius step is the long rise time. The justification for the rise time, as given by the recipe, is that this amount of time is needed for the dough to knead itself. And this does happen. Over time, the proteins that are involved in gluten formation align themselves, which will result in a chewy loaf of bread with a nice crumb. To make sure that the bread does not over-rise in this amount of time, one only adds 2.5 grams (3/8 of a teaspoon) of yeast. (A typical recipe for a loaf of bread calls for about 20 times this much yeast.) An unintended effect (at least in the original description of the recipe) is that the long rise time allows for more flavor development in the dough. Some of these flavor molecules may come from the added yeast. Other flavor molecules may come from other natural yeast and bacteria found in the flour or in the kitchen where the bread was made. Any microbe, given enough time will produce several flavor and aroma molecules from hydrated flour (dough). The bread that I make using this recipe always comes out much more flavorful than bread that I let rise over a shorter period of time (2–5 hours). Professional bakers are known to modify the flavor of their dough with pre-ferments (these can come in the form of bigas, poolishes, old dough, and sourdough starters).[5] In the no-knead bread recipe, the entirety of the dough is basically a pre-ferment.

I didn't really realize that this long ferment flavor development was going on in the dough until I read another *New York Times* article in their Food section.[6] Oliver Strand had set out to find

Pizza

the secret to making the perfect pizza. So, he went to pizza-making seminars with experts. He had gone in with the assumption that good pizza could only come from an oven that could reach intense heats. Most conventional ovens reach a maximum of 275 °C (550 °F). The pizza ovens used for Neapolitan pizzas and New York style pizzas reach twice that temperature. But, he went anyway, thinking that making good pizza at home was a lost cause. What he found was something different. The real key to good crust was using a pre-ferment. He goes on to describe how to make a sourdough starter and how to use that starter to make pizza dough. At the end of the article, he compares a pizza baked with a freshly made dough *versus* a dough that uses a pre-ferment. The pictures that accompany the article are fairly stunning. The pizza with the fresh dough has a pale color with some darker spots. The pizza made with the pre-ferment dough is a beautiful golden color, which is indicative of more Maillard reaction products and more tastiness.

Once I started making my own bread, and realized how flavorful it could be, it was only obvious that I started thinking about making pizza using it. Of course, I wasn't the first person to realize

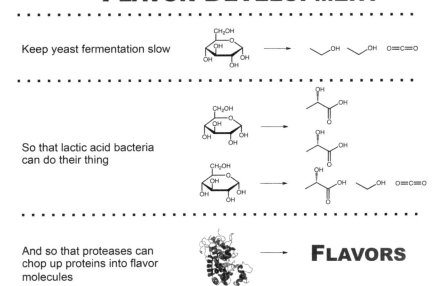

this. Jim Lahey's no-knead bread recipe sparked a thousand different variations on the Internet. People used his starting point to make almost any type of bread you could image. People also used his recipe as a starting point to make pizza dough.

My pizza dough recipe is a close approximation to the standard no-knead recipe with a couple of key differences. Michael Rulhman has written that his typical bread dough recipe is 500 g of flour, 2 teaspoons of salt, the appropriate amount of yeast, and 300 g of water.[7] For this recipe, the 5 parts flour to 3 parts water is the important thing. It gives the dough the right consistency and hydration. (The balance of the water with the other ingredients.) In all of my trials to make the right pizza dough (for my preferences), I have found that I like my pizza dough to have a slightly higher hydration percentage than my bread dough. There are a couple of reasons for this. It spreads and shapes more easily when I'm prepping the pizza. It also cooks up the way I want my crust (see earlier discussion) for the way I bake it (see the discussion later).

If I were making a New York pizza style crust, the dough would have a lower water content. There are some guidelines for using this sort of dough, though. When you start prepping this dough to make a pizza, the gluten needs a long time to relax before it can stretch far enough to make a good pizza. A brief protocol for doing this: take a rolling pin and roll out the dough, wait 5 minutes, roll it out again, wait another five minutes, repeat until the dough stretches to the length that you want with the thickness that you want.

Where I live, in Maryland near Washington, DC, there is a noticeable humidity. The 500 g of flour to 300 g of water gives me just the right hydration levels that I need for my pizza dough. In drier environments, it will be necessary to add more water. So, I put 500 grams of flour, two teaspoons of fine salt, and 3/8 of a teaspoon of dry active yeast in a plastic bucket. I add 300 g of water and mix until all of the flour is hydrated. I cover the bucket with a lid and let it sit out overnight. Two hours before I'm ready to bake the pizza I separate the dough into 4 parts, form it into balls, and let them proof. Well, that's not quite true. I save a small piece of dough and leave it in the plastic tub to mix with the next batch of dough that I make. I place the tub and the pre-ferment in the refrigerator where it will keep for a week. The more dough you make, and the longer you develop your pre-ferment, the more flavor you develop in your crust.

I can understand the thought that creating a good dough requires a lot of work. It really doesn't. A good dough requires time, a very small amount of manual labor, and a little foresight.

SAUCY

If the crust is the backbone of the pizza, and the toppings are the focal point, the sauce has to hold the whole thing together. The sauce needs to be present, but not overpowering. It needs to add lubrication, but cannot make the crust soggy. The sauce is the chef's chance to add a little acidity and bright notes to any pizza.

When I started to become interested in food chemistry, I sought out and devoured Harold McGee's *On Food and Cooking*.[8] While I've learned countless things in my multiple readings of this book, a random historical fact that McGee added surprises me every time. There are several places in his book where McGee inserts historical asides. Relevant to pizza sauce is this: Italians didn't have tomatoes and certainly didn't make tomato sauce until after Columbus returned from the Americas. There are many, including myself, who see tomatoes as inextricably linked to Italian cuisine. But that is just not the fact. The Italians (and many other Europeans) refused to eat the tomatoes that Columbus and his crew brought back. The reason for this is that tomato is in the nightshade family (along with potatoes, peppers, tobacco, belladonna, and many others). Many of these plants, specifically the ones that the Europeans were familiar with, are poisonous. So, their refusal to eat tomatoes was, at the time, justified.

Thankfully, the Italians overcame their aversion to tomatoes and figured out a way to use them to make the sauces that now go on my pizza.

There are two sauces that I consistently make for my pizzas. The first sauce is from a recipe by Marcella Hazan and is found in her book *The Essentials of Italian Cuisine*.[9] This is the best cookbook that my wife and I own. We own a lot of cookbooks, and our culinary library contains quite a few classics. No other book performs the way that Hazan's does. Every recipe that we have made from this book comes out exactly the way it supposed to, whether or not it is convoluted or simple. Her pot roast is a thing of beauty. My wife and I struggled to consistently make a tender pot roast until we tried her recipe. It comes out perfect every time. Her recipe for strawberry gelato is to die for. We make her hand-made pasta, her pesto, and an absolutely delicious brisket studded with pancetta and cooked over onions. But the gem of the entire book is also her simplest recipe.

Hazan's tomato sauce is magical. Into a saucepan add a 784 grams (1 28-ounce can) of peeled whole tomatoes, one onion (that has been cut in half), 58 grams (a half stick) of butter, and salt to taste. Cook on medium-low for about an hour, smashing the tomatoes with a wooden spoon as you go, until the sauce reaches whatever thickness it is that you would like. Remove the two onion halves. Serve. It is a foolproof recipe. Prep time is 1 minute (open a can of tomatoes, cut an onion in half, cut a stick of butter in half).

There are several small steps that you can do to make the sauce "better". The first is somewhat obvious and involves choosing the right tomatoes at the store. You want to find canned tomatoes whose flavor you like. Sadly, this takes some trial and error. There are certainly some that I really don't like. There are some that are really good.

Second, you want to look at the label on the can and look at the ingredients. Specifically, you want to see if the tomatoes contain calcium chloride. The calcium itself isn't bad or good, but it can affect the type of sauce you get. Why is there calcium added in the first place? It is there because of consumer desire and market forces. When tomatoes are canned, they are heated at high temperatures and pressures to ensure that no pathogens survive within the container. Canning is one way we make our foods shelf-safe. The side effect of this is that the tomatoes get cooked and become a little squishy. For many, this is no big deal. But, some consumers want canned tomatoes that have the consistency of regular tomatoes. Enter calcium chloride. Calcium ions have a +2 charge associated with them. Some of the cellulose-like polymers (pectin, specifically) that are naturally found in the cell walls of the tomatoes have negative charges associated with them. Calcium helps to hold the tomato cells together through the cooking process by assembling the negative charges of the polymers around its positive charges. Translation: the calcium keeps the tomatoes from getting squishy when they are canned. It has become difficult to find tomatoes without added calcium in the US. I actually prefer them without, not because I don't like the flavor, but because of how the calcium affects the sauce's consistency. In the same way that the calcium holds the tomato together during the canning process, it will hold the tomato together during sauce making. Tomatoes with added calcium will result in a sauce that is chunky (bits of tomato poking out of a thin sauce). Tomatoes without calcium, on the other hand, fall apart almost completely. The sauce that you make is smooth and uniform, naturally thickened by the hydrocolloids from the tomatoes.

The last tip is to cook the tomatoes slowly at first. Hazan notes this by saying that you should only see two or three bubbles popping up through the sauce every minute. Technically, you want to keep your sauce below 80 °C (176 °F) as you start cooking. The reason for this is biochemical. There are proteins in the tomatoes that naturally break down the cell walls. As the cell walls start to fall apart, they release hydrocolloids (cellulose, hemicellulose, pectin, and others) into the sauce. Once that happens, they hydrate and thicken the liquid. If the temperature of the sauce gets too high (above 80 °C), the cell wall-breaking proteins stop

THICKENING SAUCE

Cooking tomato sauce at a low temperature allows enzymes to chop up the pectin and cellulose in the tomato cell walls, releasing hydrocolloids into your sauce and thereby thickening it.

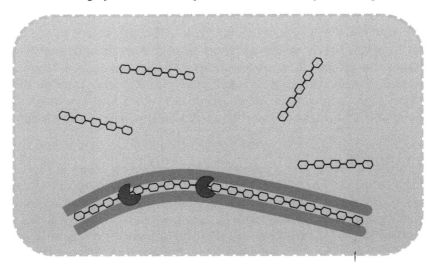

cell wall-breaking. A little patience goes a long way to making good tomato sauce.

At home, we probably make this sauce once a week. Usually, it's used for topping pasta. Every once in a while we use it on our pizza. It is always delicious. (One note for using this as pizza sauce: you want to make the sauce in advance so that it has time to cool. A warm sauce will make your dough soggy before it goes into the oven.)

The sauce that we use most of the time for pizza is from an America's Test Kitchen recipe and goes as follows:[2] to a blender add 1 can of tomatoes (just the tomatoes, not the liquid), 10 grams (1 teaspoon) of salt, 5 mL (1 teaspoon) of red wine vinegar, 4 grams (1 teaspoon) of oregano, 1 garlic clove, and 15 mL (1 tablespoon) of olive oil. Purée until smooth. Simple and easy.

When I make this sauce, I do a couple of things slightly differently. I don't add the oregano. It's not my thing, although I know some people really like it in this sauce. I use a spoon to pull out each tomato individually. While doing this, I press the tomato against the side of the can to press out any excess water. After I add the tomatoes to the blender, I let them sit, undisturbed, until they render more water. (Remember that balance we need to find with our water.) I really like this recipe for a number of reasons. The vinegar is the real secret of the whole thing. It makes the sauce and enables it to be a bright spot on the pizza. The garlic lends a natural heat, which can be intensified by the addition of pepper flakes, should you want such a thing. On that same note, it is pretty straightforward to alter the herbs and spices to match one's tastes. I remove the oregano all together. But anyone who makes this recipe can simply play around with the spice mixtures that appeal most to him or her. Finally, I don't have to cook the sauce, and I can make it in 15 minutes (with about 1 minute of effort and 14 minutes of waiting) right before I make my pizzas.

SAY CHEESE

Pizza toppings are always given top billing on the pizza. When you order a pizza, you order a pepperoni or a vegetarian or a Hawaiian or a rapini with an egg and pecorino. For any of these pizzas, cheese is really the only (nearly) omnipresent ingredient.

For my own pizzas, I use fresh mozzarella, the kind that comes shaped like a ball and stored in brine. Having covered cheese in an earlier chapter, I'm not going to dwell on it too much here. The key to a good melting cheese is that the protein that makes up the framework needs to melt at the same temperature as the milk fat that takes up residence within that structure. If the fat melts before the framework, you are left with a clumpy, oily mess. Mozzarella sits right in that sweet spot for melting with the added bonus that its calcium content continues to hold the protein fibers together even after it melts. That is, it gets gooey and stretchy.

As for the rest of my toppings, there are several things that my family and I like. Each pizza I make, no matter what else it has on it, also gets a liberal sprinkling of Parmesan cheese, which adds a little boost of flavor. Pepperoni pizzas and green olive pizzas are my kids' favorites. We make a fair number of Margarita pizzas as well. Outside of those staples, we like blackened rocket with mozzarella and shaved pecorino, with olive oil substituting for the tomato sauce. I make a prosciutto pizza that I bake in its Margarita form and add the prosciutto before slicing (baking prosciutto on top of the pizza robs it of some of its flavor and dulls the experience of eating good prosciutto). Finally, à la Zelo's pizza (see earlier), we make a blackened corn and caramelized onion pizza.

And that's the great thing about pizza; you can literally add any topping that you like. It's a pizza! Chances are, no matter what you put on it, it will still be delicious!

PIZZA TIME

Now that we've talked about all of the ingredients, it is time to make everything come together.

Again, it helps to have an idea of what you are going after (taste, texture, appearance) so that you can figure out how to get there.

In a really hot oven (475 °C), the gasses in the dough will expand before the dough has enough time to harden into a crust. This will result in a pizza with an impressive oven-spring (puffed up crust) where there are no toppings. If this is what you are going for, the dough has to be pliant (stretchable) enough to rapidly expand in the oven. This property is achieved with a dough that has a high moisture content. One problem that often arises because of the increased hydration levels and the short cooking time, is that the dough (and the ingredients) does not have the time in the oven for enough water to evaporate. When this happens, the crust in the middle of the pizza can be soggy and not maintain its shape. So, there are a couple of things to keep in mind. First, don't over stretch the dough. Dough that is too thin in the middle is more prone to being soggy. Second, keep the moisture content of your other ingredients lower (cook down or use a colander to remove excess water from the sauce, make sure that your sauce is cool when you top the dough, drain excess water from your cheese and other toppings as well).

In a conventional oven, chemical reactions don't happen quite so fast. Again, you need to find the right balance for the moisture in your ingredients. Your cheese needs to be melted and bubbly at the same time that your dough finishes cooking. For a thick dough, like you'd get in a Sicilian pie or deep dish pizza, you'll want to turn down the temperature of the oven and bake for a longer time. For a New York style pizza, you'll want to turn the oven up as high as it can go. Find a cooking time that works based on your oven. (Reminder: even if your oven tells you what the temperature is set to, that doesn't mean your oven is actually

PIZZA BAKING

at that temperature.) I have known enough people who use the America's Test Kitchen dough recipe (minimal effort here as well, the dough gets mixed in a food processor), that I often suggest this recipe.[2] Use this as your base when trying to figure out how your oven works. After you've got a good feel for things, then you can tinker around.

But, what are you to do if you want a Neapolitan style pizza at home without springing for the construction of a fire-burning oven? The key for creating a Neapolitan pizza is facilitating ultra-fast thermal transfer from the oven to the dough. At the high temperatures in wood-fired ovens, this is expected. Another kitchen gadget, one that has revolutionized the way I make my pizza, is the baking steel. Much like a baking stone, the baking steel provides a flat surface inside of your oven for making pizza and bread and other things. The difference between the stone and the steel, however, is that the steel fosters rapid energy transfer of heat from the steel to the pizza dough. This thermal conduction helps the crust to cook in mere minutes. [Neapolitan pizzas take 90 seconds to cook; the pizzas in my oven, set at 285 °C (550 °F) take 180 seconds to cook.] The problem with this method is that the crust gets cooked before the toppings are finished. Taking a note from Kenji Lopez-Alt, keeping your pizza steel on the top rack of the oven (near the broiler) and turning the broiler on when cooking the pizza, helps to solve this problem.[3]

DELICIOUS BY ANY DEFINITION

I spoke with Peter Pastan, chef and co-owner of 2 Amy's, along with Obelisk, a tremendous Italian restaurant in DC. We chatted about pizza making, dough, yeast, ovens and pizza in Naples. One of the first things I asked him was to describe his current role at 2 Amy's, to which he responded: "He who does what others won't." He seems very laid back and self-deferential with an ironic sense of humor. But this belies what comes through in our conversation, that he has an encyclopedic knowledge of Italian cuisine and the earned authority of someone who has built up this expertise over a long time. I imaging that this interplay is a result of his parents. His mother, Linda Pastan, is an author and was Poet Laureate in the state of Maryland and his father, Ira Pastan, is a scientist at the National Institutes of Health and is a pioneer of protein-receptor research. When I talked with him, they had just finished prepping dough that morning: 80 pounds of it, which would be ready for dinner service the following day. What was most fascinating to me from our conversation are the things that he admitted they have to best keep an eye on in their pizza making. He says that over the years they've got so used to making dough, that it's really become rote, as I imagine it would. The yeast, and its effects on the dough texture, really requires attention. Not only does the yeast help the dough to rise, it also breaks down some of the gluten. Use the dough too soon, and your pizza is too stiff. Use it too late, and your pizza is too flat. Yeast is a living creature, and you are always at its mercy if you want things done right. The other thing that surprised me was our conversation about oven temperature. Pastan is most concerned with the convection in their ovens. The temperature isn't as important as how heat moves through the oven. Does it whip around or is it stagnant? He tells me that he watches the flame from his fire. If there is a gentle flame coming off of the wood, the heat inside of the oven isn't working the way it's supposed to. However, if the flame is bouncing around, kissing the roof of the oven as well as the pizza, then they are doing a good job.

Pastan also told me about his preference for pizza. He'd much prefer to have his pizzas be more like what they serve in Naples, where the centers get very soggy and soupy. This is all on account of the (quality) of the ingredients that they use. The cheese is fresh and has lots of water in it, as do the tomatoes. Highlighting these ingredients, at their peak, gives you a soggier crust. Nothing you can do about it. Most American's don't want it that way though.

And, that's OK. The great thing about pizza is that it is almost always good. And, when you can make it yourself, you can have it just the way that you like it.

REFERENCES

1. M. Colonna-Dashwood and C. H. Hendon, *Water For Coffee*, Colonna-Dashwood and Hendon, Bath, 2015.
2. America's Test Kitchen, *New York Style Pizza at Home*, [document on the Internet]. Americas Test Kitchen [cited 2016

July 5], available from: https://www.americastestkitchen.com/episode/334-new-york-style-pizza-at-home.
3. K. Alt-Lopez, *The Pizza Lab*, [document on the Internet], Serious Eats [cited 2016 July 5], available from: http://slice.seriouseats.com/the_pizza_lab/.
4. M. Bittman, *The Secret of Great Bread: Let Time Do the Work*, New York Times, 2006.
5. J. Hamelman, *Bread: A Baker's Book of Techniques and Recipes*, Wiley, Hoboken, New Jersey, 2nd edn, 2013.
6. O. Strand, *The Slow Route to Homemade Pizza*, New York Times, 2010.
7. M. Ruhlman, *Ratio: the Simple Codes Behind the Craft of Everyday Cooking*, Scribner, New York City, 2009.
8. H. McGee, *On Food and Cooking*, Scribner, New York City, 2004.
9. M. Hazan, *Essentials of Classic Italian Cooking*, Alfred A. Knopf, New York City, 1992.

CHAPTER 10

Meat Time

Over the years, my wife and I have collected a number of cookbooks. I suppose that's only natural when you enjoy cooking as much as we do. I think that the first cookbook that "we" owned was a community compilation book of recipes from my then future mother-in-law; her civic group published a collection of their favorite recipes to raise money for a charity. The next cookbook I remember owning and using was one of the *Barefoot Countessa* cookbooks.[1] I think that this one came from my parents, probably as a Christmas gift. When we got married, our cookbook shelf started to turn into a cookbook library, filled mostly with classic titles like *The Joy of Cooking*.[2]

These all set us on a course for being self-reliant in the kitchen. We learned how to do some basic things. We learned how to treat ourselves every now and then. Living in Chicago on a graduate student's and non-profit worker's salaries was not conducive to dining out. Although we did take advantage of Chicago's excellent food scene (our most memorable meal was for my wife's 25th birthday at Arun's, an exquisite Thai restaurant), we needed to cook for ourselves if we wanted to spoil ourselves with a nice dinner.

Though we were becoming good cooks in our own right, I wouldn't say that we did anything too adventurous. We stuck with tried and true recipes that could nourish a couple of poor kids on a budget. Pastas, roasted chicken thighs, meals that could stretch a package of ground beef, we found a way to make good food with what we had. I remember how excited we were to find our first good produce store in Andersonville, which was an up-and-coming neighborhood in uptown Chicago at the time. The produce was both cheap and incredible. This Latino market is where we discovered the joys of mangoes and then discovered the joys of buying a full crate of mangoes at once. Exotic ingredients, things like avocados, pushed these two kids from Ohio away from their green beans, corn, and red meat roots. (It's amazing to think of produce that once seemed so strange and foreign, like avocados, that have now become a staple in our household.)

Even with access to new ingredients, our cooking didn't really travel too far outside of our comfort zone. We learned how to make guacamole. (Actually, my wife did. Her guacamole is splendiferous.) The mangoes we would just slice and serve. We might have done something as audacious as making a mango and red pepper and cilantro salsa to serve with chicken or fish.

Our first taste of cooking outside of our comfort zone came when my mother-in-law, as another Christmas gift, gave me a new cookbook, *The Zuni Café Cookbook* by Judy Rodgers.[3] The recipes in this book are, if nothing else, lyrical. Prosciutto and white rose nectarines with blanched almonds. Farrotto with dried porcini. Sand dabs with shallots, sea beans, and sherry vinegar. Guinea hen breast saltimbocca.

The food is as beautiful as the writing is engaging. But, at the time, the recipes seemed as attainable to me as that Nobel Prize-worthy work that my PhD advisor would teasingly ask me about. I had a strange relationship with this book for a while. I really wanted to make something from it. The recipes were candy for my mind. But it all seemed so distant. Many of the ingredients were foreign to me, literally and figuratively. Rodgers introduced me to the joys of making your own stocks at home. Many of these stocks were strongly suggested for other dishes in the book. Even this was intimidating. Making one thing would require making something else. This is not the way we made food at home. Preparing a meal, as defined by this book, meant something very different than my own perceptions. I struggled with her definition for a long time before I would come to understand its beauty and see its simplicity. (Certainly, Rodgers wasn't the first or only person to espouse this type of approach. But, she was the first person to do so to me. For that, I am forever changed by her and am indebted to her.) But, I didn't even know how to start with the majority of the recipes. So, I scoured the book in search of something that I felt comfortable making.

The first recipe we tried was "Pasta with braised bacon and roasted tomato sauce."

I vividly remember making this meal. Erika and I had moved to California, where I was doing my postdoctoral training at the California Institute of Technology. This was pre-kids. We were both working pretty hard. But, we had the whole Saturday. And, we made a day of it. We found a local butcher, which was much harder than I had imagined, and purchased some slab bacon. We bought the fancy San Marzano tomatoes. We bought some homemade pasta. We bought a loaf of bread and a nice bottle of wine. We took the whole day to enjoy making the sauce and having each other's company.

And the sauce did take a good part of the day. The bacon needs to be braised for over two hours in wine and stock and vermouth with *mirepoix* (onion, celery, carrot), until the bacon is "melting tender." The tomatoes get halved and placed in a baking dish with a drizzling of olive oil, conscious to not place the tomatoes too close to one another. The tomatoes are blackened over about 15 minutes in an oven set to 260 °C (500 °F). (It's best to use a ceramic dish for this. The acid and sugars in the tomatoes will ruin a metal dish. If using a metal dish, a liner of aluminum foil can save an hour of scraping later.) After both the bacon and the tomatoes are prepared, the sauce can actually get made. Soften some onion in olive oil. Add the warm tomatoes with just a bit of reserved tomato juice. Cook just long enough, Rodgers notes, "to combine the elements, but without sacrificing their textures or individuality." (I love that phrase.) In a separate pan, add the bacon, which has been sliced into 1/4 inch thick pieces, and brown. Add the bacon to the sauce and fold into cooked pasta.

I was hooked. From the time I was given the book, I knew I would love it. I just needed the courage and commitment. For my 30th birthday, Erika took me to San Francisco. Of all of the places that we could have gone in the city, one of the places we went was Zuni Café. The restaurant is tucked into a spot in San

Francisco where the commercial city center turns into industrial buildings and auto shops and then into neighborhoods. Zuni Café occupies a building that sits on the corner of two streets that come together at an acute angle. Zuni's architecture reflects this shape, narrowing to a tight corner at one end. It was the best meal we had during our visit. Most memorable to me were the small plates that we ordered. There is something luxurious about a good appetizer. To this day, when my wife and I cook and entertain, most of our time is spent on the main course. But, at some restaurants, you can tell that they put in an amount of time and effort into crafting their small plates that we still would never consider doing at home.

San Francisco holds a special place in my heart. When I went to college, I only ventured 40 minutes away from home. In the middle of my undergraduate training, I moved to Northern California to do an internship at Sandia National Laboratory in Livermore. The morning I left, my mom and I had a good cry. (I'm not the crying type until I'm a blubbery mess. There's not much in between.) I am the oldest in my family, and I don't think that either my mom or I were ready for any of her kids to really move away. My dad flew me out, always needing to be responsible for making sure we were well taken care of, and got me settled. This was the first time I was on my own. Free to my own devices. Having a working schedule, no classes and no homework, with a salary and the freedom that this allows you to explore. Being an hour away from San Francisco, I got to explore a lot.

I get to San Francisco every now and then for work. The American Chemical Society holds its bi-annual conference there every few years. It is always one of the more popular meeting sites. The convention center is right downtown. People come from all over the world to talk science and enjoy the comforts and riches of the city. Invariably, someone at the conference will ask me for suggestions of where to eat. Without hesitation, I tell them about Zuni, and I tell them they have to get two things. First among these, if they are serving it, is their ricotta gnocchi. I'm not a big fan of potato gnocchi, but the delicate and not-so-simple-to-master ricotta version is a true delight. The other dish that I tell them they have to get, and the reason why many people flock to Zuni, is for Rodgers' take on roasted chicken with bread salad.

These chickens come out of their high temperature brick oven perfectly cooked, evenly tender, no matter your preference for white or dark meat. What the meat brings in warmth and satisfaction, the skin brings in addiction. Wafer thin and crispy right out of the oven, the skin is the real reason why people order this dish, whether they admit it or not. And, in a wonderful turn of fate, anyone can make this chicken at home.

Erika and I make this roast chicken at least every other week from the start of the fall to the end of spring. Actually, that's not true. Erika makes this one. I just get to enjoy it. There are a couple of requirements needed to replicate the restaurant's version. The chicken needs to be small [between 1.25 and 1.6 kilograms (2 and 3/4 and 3 and 1/2 pounds)]. In the States, most of the chickens that you find at the grocery store are much larger, to the order of 5 pounds. So, it can take a little effort to find a store that sells chickens that are this size. But, for the way that the chicken is cooked, the size needs to be within this range for the white meat, the dark meat, and the skin to finish at the same time. The second requirement is that the chicken needs to experience some really intense heat on all sides. For making this dish, the oven is run at 246 °C (475 °F). But the chicken isn't just placed into the oven. The first step in cooking is to place the chicken in an oven-safe pan, preheated at medium high on the stovetop. Keeping the chicken in contact with a heat conductive surface during baking really helps to bring out the crispiness of the skin. The final requirement is that the chicken needs to be well salted [5 grams (3/4 of a teaspoon) per pound of chicken] 2 or 3 days before it is cooked.

In her cookbook, Rodgers reserves an entire section outlining her education and devotion to salting food early. Contrary to what she had grown up believing, that salt should be added at the last minute to avoid drying out, her culinary education in France ingrained the importance of salting for flavor and texture.

In some respects her early prejudice towards the proper way to salt is well founded. Some salting is necessary to enhance the natural flavor of the foods that we eat. Our nervous system isn't a collection of lonely circuits that run straight to our brain. They are integrated. The wires cross at different points. One impulse can completely change the way we experience a second. The way we experience flavor is no different. Within the broader scope of flavor, taste, which comes only from the receptors on our tongue (traditionally defined as salty, sweet, sour, bitter, and umami), is still difficult to fully study and understand. But, there are a few aspects of the way that we taste that scientists have developed some insight for. At the right concentrations, salt, in the food that we eat, will suppress our ability to taste bitterness and enhance our ability to taste sweet, sour, and savory flavors. Without salt, the same food will taste like a dull version of itself. Add too much salt and the food will taste salty, as you'd expect. But, with just the right amount of salt, the food can live up to its potential.

Rodgers shares some eloquent words about salting and survival; salting food, even lightly, can preserve it for a time without the need for refrigeration. This type of salting falls well short of salt curing. A little bit does go a long way. She shares an anecdote about the owner of a restaurant where she was working admonishing her to salt some sea bass they had left over from the night's service. "Salt it. It will keep perfectly. "Noting the doubt on Rodgers' face, she said, "It'll be even better." Salt has the power to do this. Like any good ally, it will help food to reach its potential without itself being noticed.

There is some experience required to know how much to salt. Seeing directions like: salt until it tastes better but without making it salty, is maddeningly vague. This is one aspect of a recipe that I still don't feel comfortable changing on the fly. Recipes in cookbooks are normally the result of unseen trial and error. The amount of salt is part of this process. Only after I have made a recipe do I feel comfortable changing this. And, if I'm just throwing a dish together on my own, I will usually seek out some guidance on salting.

The other place where Rodgers' early education in salting was correct, to a point, was in salting right before cooking. When you salt a piece of meat, the salt stays right on the edges and faces of that meat. Cooking it right away, there will be enough salt to season the meat properly. But, salting can give better results than this. As the salt sits on the meat for a while, the first place it travels to is along all of the moisture pockets between and among the muscles. In this instance, it acts as a desiccant. The salt draws the liquid away from the meat to itself; the water drawn to the attraction of the sodium ions and the chloride ions. In this case, the meat will have no choice but to be dry. The water is no longer among the muscle fibers, softening them. Its attention is only

Salting & Meat

Salting right before you cook

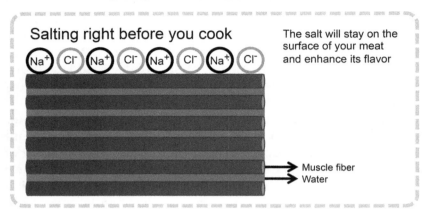

The salt will stay on the surface of your meat and enhance its flavor

→ Muscle fiber
→ Water

Salting a few hours before

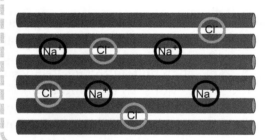

The salt will move into the water channels between the muscle fibers and draw the water to itself, making the meat taste dry.

Salting day(s) before you cook

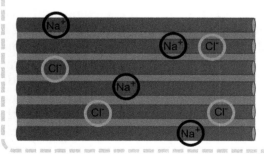

Wait longer and the salt will migrate into the muscle fibers, returning the water and disrupting protein-protein interactions, making the meat tender.

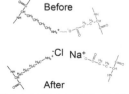

Before

After

given to the salt. In the case of the Zuni Café roasted chicken recipe, some of the salt stays on the skin and does dry it out. The ability to draw moisture out of the skin is what results in the skin getting so crispy as it cooks.

Give this seasoning a little longer to work, however, and the salt will migrate from the little pools of liquid, right into the middle of the muscle fibers. Once there, the salt can take up residence in the muscles, evenly seasoning them and making the fibers tenderer.

The chewiness or tenderness of a cut of meat is mostly defined by two variables, water and chemical bonding. The first involves how much water the muscle fibers contain. For our salted meat, the movement of the salt away from the liquid and into the muscles, frees the liquid, enhancing the tenderness. The chemical bonding part takes a slightly longer explanation. The individual proteins in muscles are held together by a mix of chemical interactions. This is a good thing. If it weren't for these interactions, our muscles would be an ineffective grouping of protein. The interconnectedness of these proteins allows them to work in concert with one another, which, in turn, allows us to move. The tightness with which these proteins hold on to one another directly influences how tender or chewy they feel in our mouths. If we can loosen these proteins, we can increase the tenderness of the meat.

One class of interactions that hold these proteins is purely electrostatic. Positive charges are attracted to negative charges. There are many of these charges on the face of a protein; positive charges coming from the amino acids arginine and lysine, and the negative charges coming from the amino acids glutamate and aspartate. As salt winds its way into the maze of muscle fibers, it will disrupt the protein–protein interactions in favor of protein–salt interactions. The positive sodium ions will interact with the negatively charged amino acids. The negative chloride ions will interact with the positively charged amino acids. The proteins don't interact with each other as much in this case because they are interacting with salt instead, leaving the meat tender.

The way that salt tenderizes meat is not a binary process. The salt needs time to diffuse through the meat. This gets to the importance of salting the Zuni chicken for 2 or 3 days. There are lots of places where salt can get slowed down on its transit from the skin to the inside of the meat. Give it time to work, and reap the benefits of flavorful and tender meat.

The debt that I most owe to Rodgers is teaching me how important of a factor time can be to cooking. Take her chicken recipe as the prime example of this. To prep the chicken before cooking, I need to salt it 2 or 3 days in advance. Now, the salting itself doesn't require a whole lot of time, maybe 10 minutes of active work at most. Taking that action, though, makes all the difference in going from good roasted chicken to great roasted chicken. Looking back at my initial reads of her cookbook, part of what intimidated me the most was the time required for all of the recipes. At the time, I couldn't imagine making my own stock over several hours, salting my meat days in advance, or any of the other time intensive techniques that she called for. But, none of these recipes require you to stare at the stove or at the meat. In many of her recipes, your active time is such a small percentage of the total time the dish needs.

Your most important tools for making any meal great are time and foresight.

This advice holds for food preparation in general, not just for salting and seasoning.

Time, in particular, holds sway over much of how we cook. Cooking is often depicted as time and temperature. Foods require cooking at a specific temperature for a certain amount of time. Taking cookies as an example. A standard chocolate chip cookie recipe calls for the cookies to be baked at 177 °C (350 °F) for 10 minutes. Increasing the oven temperature to 190 °C (375 °F) will necessitate a shorter cooking time and may result in cookies that have a very different texture. Regardless of this, we all understand how cooking makes demands of both time and temperature.

Chemistry itself is depicted in similar terms. Chemistry is thermodynamics and kinetics. A chemical reaction, the transformation of one substance into another, is defined by its thermodynamics and kinetics. Thermodynamics consist of a detailed accounting of all of the energies involved in the reaction. Some reactions require extra energy so that they can proceed; in this case the final products have a higher energy than the starting materials. Some reactions give off energy; in this case the products have a lower energy than the starting materials and the excess energy gets released to the space surrounding that reaction. Reaction kinetics involve the demands that the reaction makes on the starting materials. They must bounce into each other in just the right way, in just the right geometry, with just the right energy. In any reaction container, the starting materials are continually bouncing off of one another. It stands to reason, then, that the longer the starting materials are given to interact, the higher probability there is that they will eventually smack into each other in just the right way and will react. Further, at higher temperatures, the starting materials move around faster and will bump into each other more, resulting in a shorter amount of time needed for all of these materials react.

By these descriptions, chemistry, like cooking, is also time and temperature. Heat is one form of energy and can manifest itself as temperature. Kinetics, though they are controlled by energy and temperature, are made manifest in terms of time.

My depiction of both chemistry and cooking as being time and temperature should come as no surprise to you. This book is a culmination of those views. Cooking turns out to be just one type of chemistry. If you cook from a recipe at home, you are repeating the protocols that a chef developed through experimentation, looking for the best and most consistent results. When you modify a recipe, you are co-opting those protocols, perhaps changing an ingredient or two, looking for a slightly different reaction product. If you are introduced to a new ingredient, you analyze that ingredient before you use it. You look at its color, you feel its texture, you taste it, and you smell it.

These activities are no different than what a research or industrial chemist does on a day to day basis. We try to repeat experiments so that we can either make what another chemist has made or to become familiar with new laboratory techniques. We slightly modify experiments so that we achieve an outcome that is slightly different, to better match our needs in the lab. And, whenever we get a new chemical, the first thing that we do is to make sure that its specifications match our expectations.

In my mind, the functional difference between what we call research chemistry and what we call cooking, is the total number of variables that go into these crafts. In chemistry, such a high level of precision is required that we demand a high level of control over experimental variables. To realize this precision, we remove as many variables as we can. This is not the case for cooking. In cooking, we don't nearly have the amount of control that we believe we have. Oven temperatures are usually not the same as what you set them to. Measuring spoons don't always add the amounts that you would expect. The size of different ingredients is hugely variable (just think about different egg sizes.) In cooking, we often need to develop recipes that can tolerate these broad variations.

But the biggest difference between research chemistry (both laboratory and industrial) and cooking involves the number of reactions that are being controlled at once. In a chemical process that I study in the laboratory or that one of my peers runs in a factory, there is usually only one reaction happening at a time. When I am cooking a dish at home, there can be tens or even hundreds of reactions that are going on in a single pot. When I think about cooking in terms of the different reactions that are occurring and compare it to the research that I do in my lab, my research seems like child's play.

Much of cooking, then, turns out to hinge on finding the right time and temperature so that all of these (tens or hundreds of) reactions reach their ideal conditions at the exact same time.

Steak is a lovely example as this. Cooking a steak requires very little effort and very few ingredients (usually just a steak and some salt). But how do you prepare a steak? Let's say that you like your steaks cooked to medium. Medium corresponds to an internal temperature of roughly 60 °C (140 °F). There are a number of ways to reach this temperature using standard kitchen equipment. (Assume the appropriate amount of time for each method.) You could put your steak in boiling water. You could bake it in the oven. You could fry it in a pan. Or you could cook it on a grill.

Now, I don't know about you, but a steak that has been boiled or baked does not sound all that appetizing to me.

Going back to the point that I made earlier, if there were only a single chemical reaction that was contained within a recipe, we could cook steaks this way. The fact of the matter is though, that we need to conduct an orchestra of reactions to take our food from limpid to lyrical.

This is no different when making steak, which involves two classes of reactions: protein unfolding and the Maillard reaction.

Steak, like other cuts of meat, just comes from the muscles of some animal. As teenagers, when we start to learn more in-depth about biology and anatomy, two of the first proteins we learn about are actin and myosin. Actin and myosin are the two proteins that make up most of our muscles and are responsible for our movement. Actin is reminiscent of a chain of beads, the chain being opened and stretched out from end to end. Myosin hangs out on a string of its own. But, instead of a lot of beads bunched together, there is plenty of space between each myosin protein, giving them room to move. Myosin is more reminiscent of a headless person, feet firmly planted on its string and arms reaching outwards. In our muscles, the string of myosin orients itself so that the protein "hands" reach up towards a string of actin. Our muscles contract when one set of hands grabs onto actin and passes it down to the next set of hands, which can do the same.

When we cook meat, there is a noticeable change in texture. The longer we heat that meat, and as that meat reaches higher and higher temperatures, the meat becomes firmer. Much like cooked eggs, the change in texture is due to proteins unfolding and aggregating. As actin and myosin are the primary proteins making up muscles, it only make sense that they would also be responsible for the consistency of meat as it cooks. Furthermore, the differences between the two proteins are responsible for why meat can be cooked to medium or well done. Myosin unfolding becomes noticeable at 50 °C (122 °F).[4] Actin unfolding becomes noticeable at 66 °C (150 °F). A well-done steak is one that has seen all of its actin unfold (by 74 °C), causing it to become tougher. The variation then, between medium-rare (54 °C) and medium-well (65 °C), is mostly due to the muscle fibers drawing closer together, pushing out water. Kenji Lopez-Alt, cookbook author and purveyor of kitchen science, likens this to squeezing toothpaste out of a tube.[5] In these descriptions, the temperatures correspond to the reaction thermodynamics. Myosin requires a temperature of 50 °C before it will unfold.[4] Actin requires a temperature of 66 °C before it will unfold. Said in a slightly different way, myosin will keep its structure at temperatures below 50 °C, and actin will retain its structure at temperatures below 66 °C.[4]

Protein unfolding isn't really a binary process where all of the proteins go from a single folded structure to a completely

Time & Temperature

Cooking meat involves unfolding proteins

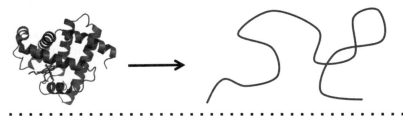

If you don't add enough energy (temperature), it won't unfold no matter how long you wait.

Unfolding at a lower energy (lower temperature) takes longer than …

Unfolding at a higher energy (higher temperature)

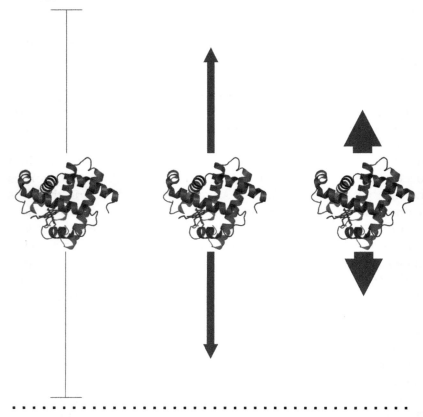

unraveled form with a flick of a switch. There are steps by which proteins lose their structure. Some steps are bigger than others and take more time. Additionally, all of the actin proteins in the chain don't all take the same steps at the same time. There may be some in the chain that are unfolded, while some remain folded. In the context of research science, we talk about these in terms of populations. There is a population of proteins that remain folded and a population of proteins that are unfolded. (Just like there is a population of young parents who drive minivans and a population of parents who haven't given in to that temptation yet.)

The unraveling of myosin is more complex than the unraveling of actin.[4] The "hands" of myosin are the first parts to go, and at 47 °C, there is some small population of myosin "hands" that are unfolded. By 55 °C, the hands have lost all of their original structure. The next part to go is myosin's "body". It starts to go flabby at 56 °C and has completely let itself go by 58 °C. The population shifts for actin start at 63 °C and are finished by 66 °C.

The kinetics of these unfolding processes don't really come into play when you are cooking steaks. That is, if your steak reaches 55 °C, myosin will unfold quickly enough that you don't have to worry about the amount of time that the steak stays at 55 °C. You don't have to wait any longer than it takes to move your steak from the pan to the plate.

The way that time matters for cooking steak comes about in a more subversive, yet completely expected way. Once you place a steak onto a pan, it does take time for its internal temperature to come up to 55 °C. Much like when you put a pot of water on the stove to boil, it doesn't happen right away. The higher you have the stove turned up, the faster it will happen. There are equations of heat transfer that model how energy moves from a heat source through a piece of meat. Peter Barham expertly engages readers with a full description of these in his book, *The Science of Cooking*.[6] While this book leans heavily towards chemistry, his book is an ode to the physics of cooking. One of my favorite parts of his book involves the physics of heat transfer in roasting a chicken or turkey. Heating rates are even (and more easily predicted) for spherical shapes. Trussed poultry doesn't quite approximate a sphere, but it does so more closely than the spherical cows that show up in the fields of a physicist's mind.[7] This leads to complications that arise when roasting a chicken that are compounded by needing to cook the thighs and legs faster than the breasts. His discussion is a fun exercise of how physics and math can model how a chicken is roasted.

If cooking a steak were just a matter of picking which proteins to unfold, it would be really easy. But a good steak doesn't have just a nice pink inside, it has a surface that brown and crisp and full of flavor. The Maillard reaction provides the unmistakable burst of flavor that we have come to expect from the best restaurants. The golden color and savory crunch are what set steaks apart.

This perfectly shows how chefs have to balance multiple chemical processes to achieve a desired result for a dish. The steak needs to reach the right internal temperature at the same time that the Maillard reaction reaches perfection. (Another note that the Maillard reaction is not a single reaction. It is a collection of hundreds of complex chemical processes.) The need to balance many reactions is also why there are some cooking methods that are inappropriate for cooking steaks. In boiling water, a steak may reach the right internal temperature quickly, but there is no chance that the Maillard reaction will happen in that amount of time in that particular environment. The result is a gray and limpid piece of meat; the change in color a result of the protein myoglobin leaching from the muscle. In an oven set at, say, 190 °C (375 °F), the steak may well see a little Maillard going on at its surface over the longer time it takes to cook. However, cutting into the steak, we will find another problem. The very middle of the steak may be the right color of red and just the right doneness. But, all around that will be meat that is overcooked and dry. The physics of heat transfer in an oven are just too slow to properly cook a steak.

In a hot pan or on a grill, the physics and chemistry start to work together to our advantage. The searing heat of the pan focuses its efforts on the surface of the steak. The Maillard reaction gets carried out rapidly, with a hotter surface browning faster than a cooler surface. The heat intensity at the surface decays as heat moves through the meat. The heating of the inside of this steak is fairly even, giving you a broad swath of reddish-pink in the middle, if that's what you're looking for. Cooking a steak becomes a big game of related rates (raise your hand if you remember doing

COOKING MEAT

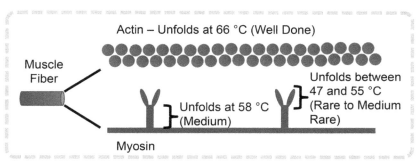

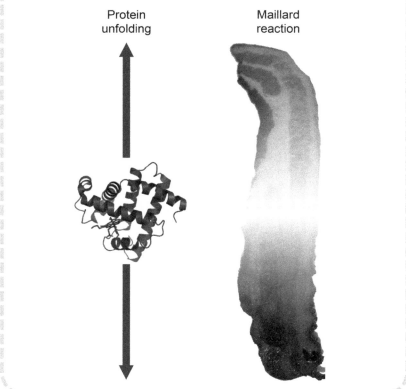

related rate problems in calculus as a teenager). That is, what stove-top setting is required to get the inside the right doneness at the same time as the Maillard reaction gets the crust where you want it.

At higher temperatures, the Maillard reaction is the most finicky. There is a fine line between perfectly browned and blackened. A steak can go from savory to acrid in the blink of an eye.

Harold McGee has devised his own way for cooking steaks that is slightly different from conventional methods. When I cook a steak at home, I sear the steak on one side over about 4 minutes or so. During this time, the steak develops a crust and starts to warm the inside. After this time, I flip the steak and repeat for the other side. Conventional wisdom held is that you should not disturb the steak during this period as you won't develop a crust that is nearly as nice. McGee, however, showed that when you flip your steak every 30 seconds, you are sill able to get the crust that you want and the insides get cooked more evenly.[8]

Another way to cook a steak, one that seems more like laboratory chemistry than cooking, is *sous vide*. The rough translation for *sous vide*, under vacuum, doesn't really get at the heart of how this technique works. The power of *sous vide* cooking comes from an exacting control over temperature. By setting a water bath to a specific temperature, you can ensure that your steak is never overdone. That is, if you want a steak cooked to 55 °C, set the water bath to 55 °C. The name *sous vide*, then, comes from the fact that the foods you cook this way are placed in vacuum sealed bags. These bags keep the water in the bath from ruining your food. The air gets sucked out of the bags to increase the thermal contact between the water and the food, shortening your cooking time. One of the drawbacks to cooking this way is that you need time, over an hour for your steak to come up to temperature.

Again, *sous vide* cooking is something more akin to laboratory chemistry because you decouple the protein unfolding processes from the Maillard reaction. Isolated, you can run each of them on their own. I prefer to sear my steak after *sous vide*. This is called reverse-searing. Slowly cook your steak until done and then put a crust on it. A steak just out of a *sous vide* water bath will be pre warmed. After a quick pat down with a paper towel (water will steam as soon as it hits the pan and slow down the Maillard reaction), the steak is placed in a raging hot pan and quickly browned.

The short amount of time in the pan ensures that the proteins in the steak don't get cooked much more than they already are. Preparing steaks this way results in a nicely crusted, perfectly done steak (cooked to your specification) that maintains a single color throughout. There is no gradient of doneness. If you want a pink steak, the entire steak, top to bottom, will be pink save for the very outside where a savory and crunchy Maillard wrapper enhances your eating experience.

While *sous vide* steaks at home can really enhance your dining experience, they can still fail to live up to the best steak at a quality restaurant. The easiest way to boost flavor is by adding butter. Make sure that you baste your steak in butter before you sear it. Then pour butter on it. Then serve it with some melted butter. The fact that butter gets used in professional kitchens so often speaks to a love that many of us won't admit.

There is another advantage that most restaurants have over us who toil away in our unprofessional kitchens.

Heston Blumenthal is often recognized as one of the best chefs in the United Kingdom and one of the most forward thinking chefs in the world in his unabashed use of the most obscure or innovative technology in preparing his dishes. Together with the BBC he made a television series titled, "Heston Blumenthal: In Search of Total Perfection."[9] This phrase played on the exacting requirements he demands of the food made at his restaurants. Blumenthal openly acknowledges that perfection is a subjective description. The point of the series, which was also turned into a book by the same name, is that Blumenthal wanted to learn from diverse sources ranging from chefs that are recognized as the best in the world, to amateur cooks who have toiled for years on their own to craft singular dishes. In understanding what made particular dishes singular, he applied what he learned in an effort to create his own dishes that were faithful interpretations of what he saw as their most identifying marks.

For the segment on steak, Blumenthal travelled to New York City. Upon visiting the Penthouse Executive Club, the head chef, Adam Perry Lang, took Blumenthal into the restaurant's meat ager. Blumenthal writes, "The unusual hues and still-life formality made it look like some kind of weird art installation, but it was the smell that made the biggest impression." He goes on to describe the aroma as "nutty and grassy, with a strong blue-cheese note."

In dry aging, meat is stored in controlled conditions (temperature and humidity) and is left to its own devices. Many of us forget that meat comes from living creatures, walking biology. Biology doesn't end when an animal is butchered. There are many proteins that continue to do their job well after, and completely unaware, that the animal it had once been a part of has ceased to exist. For the dry-aging process, the work of proteases is highlighted. Biology is the ultimate repurposer of materials, recycling tissue so that it can be rebuilt or reimagined. Muscles are not lacking of proteases, specialized protein enzymes whose sole job is to break down old and worn proteins, such as the protein found in muscle tissue. During the dry aging process, these proteases chop up the proteins found in muscle fibers. This has a number of effects. The meat becomes tenderized and the muscle fibers get broken down. The result of breaking down the meat is the presence of amino acids, which have a savory flavor when they are on their own. The prototypical example of this is glutamate, of monosodium glutamate (or MSG) fame. Other enzymes present in the muscles break down fat molecules into the more funky aromas that are most noticeable in dry-aged meat.

Blumenthal took these lessons to heart. He wrote about the state of beef production in the UK and what it can do to reach the quality found in the States. But, mostly, he wrote about the qualities that set dry-aged beef apart from its un-aged cousin: tenderness, savoriness, and funk. Being true to the type of chef that he is, he went about finding different ways to recreate these properties. By heating the food for a long time at temperatures just under 50 °C for 18 hours, he attempted to highjack and kick-start the muscle fiber-chopping ability of different proteases, calpains and cathepsins. He found a way to amplify the funkiness and savoriness by making a blue cheese-infused butter and a mushroom ketchup.

Here in the States, dry-aged steaks aren't really a secret. Many restaurants market the fact that they use dry-aged steaks. And, for a serious premium, there are many butchers that will sell you dry-aged steaks. Of course that doesn't stop some people from trying to do it at home. On his blog, Kenji Lopez-Alt has a collection of posts that he has titled, "The Food Lab's Complete Guide to Dry-Aging Beef at Home."[10] He describes how to make your own set up using a mini-refrigerator modified to couple to some sort of humidity control. Kenji also guides potential users on how to age their steaks and what they can expect over time. He found that most of his taste testers (really ... how can I get this job) preferred their steaks aged to 45 days, at which point the funkiness was noticeable but not overpowering.

IT'S ABOUT TIME

Any way you slice it, it seems that most meat can benefit from a careful application of time. And, while most of us will never age our own steaks for over a month, planning and patience are often the best ingredients for making any kind of meat.

REFERENCES

1. I. Garten, and M. Stewart, *The Barefoot Contessa Cookbook*, Clarkson Potter, New York City, 1999.
2. I. Bombauer, M. R. Becker and E. Becker, *The New All Purpose: Joy of Cooking*, Scribner, New York City, 1997.
3. J. Rodgers, *The Zuni Café Cookbook: A Compendium of Recipes and Cooking Lessons from San Francisco's Beloved Restaurant*, W.W. Norton, New York City, 2002.
4. T. Dergez, D. Lorinczy, F. Könczöl, N. Farkas and J. Belagyi, Differential Scanning Calorimetry Study of Glycerinated Rabbit Psoas Muscle Fibers in Intermediate State of ATP Hydrolysis, *BMC Struct. Biol.*, 2007, 7(41), 1.
5. K. Lopez-Alt, *The Food Lab*, W. W. Norton, New York City, 2015.
6. P. Barham, *The Science of Cooking*, Springer, Berlin, 2001.
7. R. Allain, *What is up with the Spherical Cow?*, Wired, 2011.
8. F. Cloake, *How to Cook the Perfect Steak* [document on the Internet], The Guardian, 2012 October 24 [cited 2016 July 5], available from: https://www.theguardian.com/lifeandstyle/wordofmouth/2012/oct/25/how-to-cook-the-perfect-steak.
9. H. Blumenthal, in *Search of Total Perfection*, Bloomsbury, London, 2006.
10. K. Lopez-Alt, *The Food Lab's Complete Guide to Dry-Aging Beef at Home* [document on the Internet], Serious Eats [cited 2016 July 5], available from: http://www.seriouseats.com/2013/03/the-food-lab-complete-guide-to-dry-aging-beef-at-home.html.

CHAPTER 11

More Meat Time

There are many pleasures to be had from eating food. It can be nourishing. It can be comforting. It can be celebratory. It can be a distraction from other things in your life. (I know a lot of people who turn to ice cream when they're feeling down. My meal of choice when I'm frustrated or anxious or over-wrought is any sort of American-ized Mexican food.) But, sometimes we eat meals that completely catch us off guard. A good meal can upend our notions of what a dish should be and how it should be cooked.

There are many people who seek out mind-changing meals. Some of the most famous chefs in the world became that way because of the playful approach they take, exploiting technology, to serve meals that engage and delight. Ferran Adrià is considered the most prominent pioneer of this movement. Adrià's restaurant elBulli became an international destination. Other chefs who carry this same mantle include Heston Blumenthal (Fat Duck Restaurant, Bray, Berkshire, England), Wylie Dufresne (formerly of wd ~ 50, New York City, USA), Grant Achatz (Alinea, Chicago, Illinois, USA), and José Andrés (minibar, Washington, DC, USA) among many others.

Mind-bending meals don't have to be foods that have never been imagined. They often involve foods that we have become so familiar with that they resonate with emotional meaning. In

---

Chemistry in Your Kitchen
By Matthew Hartings
© Matthew Hartings, 2017
Published by the Royal Society of Chemistry, www.rsc.org

the movie *Ratatouille*, the culminating scene involves the preparation of the eponymous dish.[1] One of the diners immediately disregards it as a peasant's dish. But, this contrast (an unassuming meal prepared with care and precision and fresh eyes toward *haute cuisine*) heightens the enjoyment of another diner, who happens to be a food critic. As viewers, we are treated to that critic's flashback to his childhood, recalling the essence of *ratatouille* as comfort and comforting food.

I have been fortunate to have had a number of these experiences in my life. The most memorable dish I have ever eaten was at Arun's, a highly regarded Thai restaurant in Chicago. Erika and I were there to celebrate her 25th birthday. We had never eaten from a *prix fixe* menu before and were a little anxious about the situation. Would we know how to act? Do we just sit there and wait until food started coming out? Could we really afford this? Eventually we relaxed. It was difficult not to with such luxurious food. I particularly remember an exquisite piece of pike with scorched skin served with a rhubarb sauce. But the dish that most stood out to me at the time, and that I still recall to this day as the most interesting thing I have ever eaten, was served on a leaf. Our server brought out a small plate that had a leaf, which was roughly the width of a coffee cup. The leaf was piled with several ingredients: vegetables, herbs, peanuts, and a sliver of pepper. Each bite was different than the one before it. Each bite was consistently incredible.

Another meal that stands out to me, but for a different reason, is one that completely changed the way I think about food and cooking. The meal was pot roast, and it was served at a party at the Athenaeum on the campus of the California Institute of Technology.

The Athenaeum is the faculty club at Caltech. (The building makes a cameo in *Beverly Hills Cop* in a scene where Eddie Murphy's character, Axle Foley, tosses one of the bad guys through a brunch buffet.[2]) I have never really been anywhere quite like it. It is part fine dining and refinery and part unassuming and welcoming. The place is such an institution that even the graduate students and postdoctoral fellows pay for yearly memberships. In the summer months, which in southern California stretches from March through October, the Athenaeum sets up an open-air bar and grill. The outdoor dining area might be one of my favorite places in all of California. On Friday nights everyone would go to the Ath. My coworkers would have had just put in a strenuous week in the lab. There was nothing quite so refreshing as sitting out in the late afternoon sun, eating popcorn and drinking beer. Life was good at the Ath.

I didn't get to take advantage of the fine dining option at the Ath as often. It was more costly and didn't exude the laid back easiness of the outdoor option. When I did go there, it was usually on someone else's tab. My lab-mates and I would take seminar speakers to lunch, paid for by the department. We would bring prospective students, paid for by the department. We would go to fancy awards banquets there, paid for by the department and, in part, by my boss, Harry Gray.

I talk about Harry a bit more in the chapter on color, but, pertinent to this story are Harry's generosity, his joy in hosting people, and his stature in the field of chemistry; which all culminated in Harry receiving a lots of awards and throwing parties to celebrate.

On one particular occasion, it seemed as though Harry had invited all of Pasadena to his party at the Athenaeum. I have never seen the place more filled with tables or people. Celebratory banquets always included plenty of wine and passed appetizers. Dinner that night was a showstopper. Roast beef.

This roast beef was something I never knew it could be. Red on the inside. Cooked to medium. Charred on the outside. Melt-in-your-mouth good.

I was confused.

I have been eating roast beef for most of my life. Roasts are cheaper by the pound than other cuts of meat. Roasts can be stretched out, serving them with root vegetables along with either egg noodles or mashed potatoes. A single roast can provide quite a few meals.

When I was a kid, we raised a few calves on our land. Brownie and Clownie were our first two, raised and butchered and stored away in our chest freezer. One of my parents' jokes was that some nights we would eat brownies with Brownie for dinner. At any rate, our freezer always had plenty of meat in it, mostly roasts and ground beef; meat that could be stretched out to feed a growing family.

When my siblings and I were little, dinner was never a culinary event. Don't get me wrong; the food was always good. But, dinner was as much about just getting food on the table as it was getting us balanced nutrition. Both of my parents worked. Dad commuted an hour each way every day for his job in financial planning. Mom was a science teacher (physics and chemistry) at the local high school. She was in charge of making dinners for us. Because she was so busy with work and making sure that we were taken care of, our crockpot got a lot of use. Pot roasts, for her, were a perfect crock-pot meal. She would throw some roast in the crock-pot with a can of cream of mushroom soup and some carrots and potatoes, turn the crock-pot on low, go to work, come back home, and dinner would be ready and waiting.

Mixed amidst the simple things I remember about childhood, is the memory of my dad spreading the sauce from the bottom of the crock-pot all over toast.

There was nothing fancy about this pot roast. That didn't mean it was lacking in its flavor or ability to comfort on chilly nights.

Though Erika had a different upbringing than I had, pot roast played an essential part in her upbringing as well. As opposed to my family, hers had more French influence showing up in their cooking. My mother-in-law would coat the roast in a mix of cornstarch and salt and pepper, brown the meat on all sides, and braise it in a wine and stock-based sauce.

Roast beef has always been a favorite of both Erika and me; emotional connections to iconic childhood meals run strong. It was essential that we be able to make this meal on our own. I think that every young couple goes through struggles that they look on later in life and find funny. We had several notable disasters with pot roast. After a series of fits and starts, my wife and I finally found a recipe that works every time. The pot roast recipe in Marcella Hazan's *Essentials of Classic Italian Cooking* has never let us down.[3] (Our take on this, which includes only slight variations involves the following. A pot roast is browned on all sides in a Dutch oven. The meat is set aside on a plate. More oil and some butter are added along with a diced onion, which is allowed to soften. The pot is deglazed with a 360 mL (1 1/2 cups) of red wine. The meat and 240 mL (1 cup) of beef broth is added along with a single canned plum tomato. The pot is covered with a lid and placed in an oven set to 177 °C (350 °F). The roast is flipped in the pan once every 30 minutes until the meat falls apart when prodded with a fork.)

The best roast beef meals, no matter how they are cooked, all have several things in common. They are well browned and savory. They are very filling. And, perfectly done, the meat falls apart on your fork; it is soft and hearty all at once.

Given what these cuts of meat are like before they are cooked, that the end result should be so tender is completely surprising. A pot roast starts out as a really tough piece of meat. Coming from the front shoulder of a cow, the pot roast is taken from a muscle that is responsible for supporting the animal, while it stands, and for enabling it to amble about. If you try to cook it like a steak, you'll end up with something that eats like a rubber tire.

Your approach when you cook a roast, *versus* your approach for cooking a steak, are necessarily different from one another. A steak, starting off tender, stays tender when cooking over extreme heat for a short amount of time. A roast requires cooking at a moderate heat for a long period of time so that it can transition from tough to tender. A typical roast recipe will call for a chef to cook the meat for 3 hours at 177 °C (350 °F). If you tried to cook a piece of filet the same way, you would be left with a hockey puck at the end of the 3 hours.

All muscles rely on actin and myosin for movement. Filets and pot roasts are no different, in this respect. The reason why these two cuts of meat are so different is, in part, due to what each of these muscles are used for. The filet and other steaks come from the mid-section of the cow. The cuts that come from the legs and shoulders and hips of the cow are different. These muscles are constantly being used for motion or support. And, because some muscles need more help than others, nature has devised a different protein to support the muscles that most support us.

There are scores of different proteins in our bodies. Each of these has a different job that enables us to live the lives that we do. Of all of the proteins in our body, none are used as much as collagen.[4] Collagen accounts for around 30% of the total protein content in a person. Our cartilage and tendons and ligaments are mostly just collagen. Our bones, although we think of them as being calcium, are mostly collagen. Our organs are mostly put together with collagen. And, importantly to this conversation, collagen is used to hold individual muscle fibers together. To prevent them from fraying and becoming incapacitated, collagen provides a protective sheath for the most vulnerable and active muscle fibers.

From where we find collagen, we can surmise how nature uses it: to give structure and shape to our bodies and to provide support where needed. To maintain the strength and mechanical properties that nature demands of it, collagen winds itself into a very unique shape. Like many other biological molecules, collagen adopts a helical structure; a shape that looks like a spring that was stretched too far. The helix, of course, is familiar to many because of the iconic double helix of DNA. In this shape, each of the coils follows the twisting of the other, wrapped in a tight embrace. Each of these coils works in unison to bring stability to the entire molecule.

Collagen one-ups DNA. It is a triple helix. Three individual strands of proteins, each adopting a helical structure on its own, twist around each other.

By way of an outdated analogy, I have a phone on my desk that sports a long black cord that runs from the body to the handset. The cord itself is neatly coiled. The coil allows the cord to extend and bend when needed. The elasticity, its ability to extend and return to its original shape, adds another layer of strength while keeping it as compact as possible. Now, my particular phone cord is tricky. It is long enough and the coils are just loose enough that it is always getting tangled. My tangled cord means that I have to sit by my desk at really awkward angles whenever someone calls. The way the coils interact make it much more difficult to move the handset from the base. Collagen is kind of like that. Three coils, fitting together snugly, make for a much stronger cord than a single coil by itself.

This is all well and good for living creatures. But how do you go about cooking collagen? How do you soften it up so that its easier to digest?

For this approach, we can think of collagen just like all of the other proteins we have discussed in this book. To tenderize it, it needs to be unraveled and untwisted. Collagen is so stable that the defining characteristic of its unfolding is just how long it takes to accomplish this feat. (I have personally only unwound my telephone cord 3 times in the past 6 years. It just takes too darn long, and I really don't want to sit there and do it.) Once it does unravel, though, collagen turns into something magical that benefits all sorts of foods, not just roasted meats.

After it unfolds, collagen becomes partially hydrolyzed. This is fancy chemist-speak for saying that the protein is chopped apart into smaller pieces. We have a culinary term that describes these little chunks of collagen: gelatin. Gelatin, the same stuff that turns flavored water into Jell-o, plays a number of other roles in the cooking world. Gelatin can be used to thicken ice creams, ensuring that ice formation is kept to a minimum. Gelatin gets used in the production of gummy bears and jellybeans, probably its most noble use. And, finally the gelatin sets from hydrolyzed collagen after you pull a pot roast out of the oven, completing the transition from tough and chewy to tender and juicy.

When you throw a pot roast into the oven there are a couple of things that happen. Just like that filet, the heat in the oven goes to work on actin and myosin. Myosin unfolds first, as soon as the meat hits 50 °C (122 °F). Actin is the second to go all squishy, losing its structure when the temperature climbs to 66 °C (150 °F). At this point, your pot roast has been cooked to well done. This is how most of us think about our pot roasts, dark brown, just like an overcooked (to me) steak.

COLLAGEN

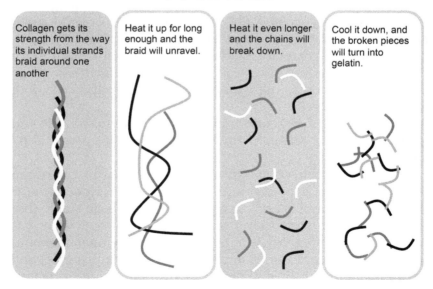

But, at this point, your roast isn't done. It's still going to be tough, tougher than a well done steak. All of that collagen is still doing what nature intended: holding muscle fibers together. But during all of that time in the oven, as the temperature is pushing forward, the collagen starts slowly and quietly unraveling.

This really is a process that you have to be patient with. When you cook a steak, you only have to cook to a temperature to get the doneness that you want. For steaks, perfection comes from a heavy, hot pan and is found out by an instant read thermometer. My own steaks have gotten remarkably better since I started using a thermometer. As a family, the type of steak we have most often is hangar steak. It pulls in flavor in a really delightful way. But, it cooks just a little bit differently from a filet or a New York Strip or a ribeye, which is my absolute favorite. The thermometer is my crutch (a crutch that is really easy to use) and allows me to get my steaks done perfectly.

But, when you are making a roast in the oven, temperature doesn't really do it for you. There are guidelines that you can use. You can cook your roast to an internal temperature of 85 to 99 °C (185 to 210 °F), pull the roast out of the oven, and keep it covered on the stove for a bit, while the collagen is turning into gelatin. I cook my roasts by feel. I want them to be fork-tender when I eat them, so I cook them to fork-tender. Some people cook their steaks by feel, testing how firm they are as they cook. For me though, the results I get from doing this don't always meet my expectations. Accuracy for steaks comes from temperature. Accuracy for roasts comes from feel.

And the payoff is worth it. Nothing is as welcoming or satisfying on a cold day than a good roast.

All of this added up to my confusion about the roast I had at the Ath, which was cooked to medium.

Every recipe for every pot roast that I had ever eaten necessarily brings the meat to well done. Baking at 177 °C (350 °F) for 3 hours will unravel both myosin and actin. In meat cooked to medium, actin doesn't get touched. Meat cooked to medium does not get above 63 °C (145 °F).

And that gives us a clue about what's going on with collagen. There are characteristic unfolding temperatures for all proteins. Just for the one's we've talked about: myosin unfolds at 50 °C, and actin unfolds at 66 °C. Give a protein enough energy and it will find a way to wiggle out of the shape it's normally in. Collagen is no different. Let's use the analogy of the phone cord to think about the major proteins found in meat. Myosin goes the easiest. If I bring my 5-year old to my office, he would be able to pull a myosin phone cord apart, and he'd be able to do it quickly. Actin is the most difficult of the bunch. Jack couldn't do it on his own, nor could I. I could unravel an actin phone cord if I went and got help from one of my coworkers. With the two of us working, we'd make quick work of the actin. The collagen phone cord, I could do by myself. I'd have to pick and poke and prod for a long time, but I could do it. With a little help from my coworker, we could make quicker work of pulling apart a collagen phone cord. But it would still take longer than the myosin and actin versions.

Collagen will unfold at 60 °C (140 °F) if you give it enough time.[5] For collagen enough time turns out to be somewhere between 1 and 2 days. And that is the secret for keeping your roast at medium. Even with all of that time, it never gets the energy to unfold actin and toughen the meat. So, you get the added benefit of collagen softening the roast along with keeping the meat from cooking to well done.

Retaining the red color in meat also comes with keeping the temperature low. That color comes from a protein named myoglobin. Myoglobin is just like any other protein (it consists of a long chain of amino acids that folds into a compact 3-dimensional structure) with one minor difference. Buried in the middle of myoglobin is a small iron complex called a heme group. You can find this same heme prosthetic in a number of proteins, including hemoglobin, as some of you might have been able to guess. The purpose of the iron atom in both myoglobin and hemoglobin, is to allow the protein to latch onto oxygen. Hemoglobin, in our bloodstream, nabs oxygen from our lungs and carries it around to whatever parts of our bodies need it. When hemoglobin reaches our muscles, it passes the oxygen off to myoglobin. While hemoglobin needs to be good at bringing oxygen from one place to another, myoglobin needs to be able to hold on to that oxygen for as long as possible. The red color of our muscles (and of red meat) is there because oxygen is attached to the iron atom, which is attached to the heme group, which is sandwiched in the middle of the protein. Change any of these variables, and the color of myoglobin will change. Myoglobin, much like collagen, unravels at just over 60 °C (140 °F) but does so much faster.

MYOGLOBIN

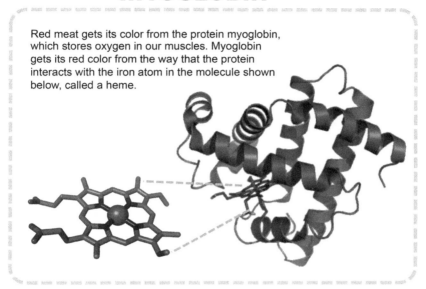

Red meat gets its color from the protein myoglobin, which stores oxygen in our muscles. Myoglobin gets its red color from the way that the protein interacts with the iron atom in the molecule shown below, called a heme.

If myoglobin wiggles open, the color of your meat will change from red to brown and eventually to gray. If you can keep the temperature below 140 °F, myoglobin, and your meat, will keep its red color.

There are a couple of ways that a chef can cook a roast to medium. You can season and wrap the meat in a vacuum-sealed bag and using a *sous vide* water bath to cook the roast. There are also some ovens that will consistently hold a low temperature, although these are more difficult to come by outside of professional kitchens.

(Whenever you cook this way, really low and really slow, you need to be cognizant about the quality of the meat that you are using. At these low temperatures (and over the long cooking times) there is a lot of opportunity for pathogens (especially bacteria) to grow on the surface of the meat. There is a real health hazard to cooking this way and any chef who uses these methods needs to be aware of the different ways to minimize the risk of getting sick.)

The magic of collagen is too attractive to pass up. While a good steak, cooked to medium, can exude tender extravagance, it is worth the extra effort to work collagen into submission. Other

SMOKING/ROASTING

common high collagen foods include ribs, briskets, pork shoulder, and poultry thighs and legs.

The other cooking technique, which slowly works at collagen but I have yet to mention, is smoking. Although smoking often cooks meat at temperatures closer to 160 °F, the slow infusion of flavor and slow infusion of collagen and the slow development of the Maillard reaction, all come together perfectly to cook high collagen foods.

Of all of the food science reference books that have been written, none is more complete or thorough or beautiful, than Nathan Myhrvold's *Modernist Cuisine*.[6] The amount of information in this 5-volume set (6 volumes if you count the recipe collection) is staggering. Because I have been interested in food science for a long time, I can't say that there is much information in the book that really surprised me. I have learned a great deal, though, especially about some of the more modernist or molecular gastronomy techniques. But, I was legitimately astonished as I read the section on smoking. There are so many intellectual nuggets tucked away in these chapters that I just didn't know before reading it. One tidbit is that the ideal coal or wood temperature for smoking is 400 °C (750 °F). That doesn't mean that the internal temperature of the smoker reaches that temperature. But, when wood breaks down at this temperature, it goes through a process called pyrolysis without actually burning. These temperatures are where the aromas effusing from the wood are at their optimal. Spicy (guaiacols), clove (isoeugenol), smoky (creosols), vanilla (vanillin), caramel (furans), and nuttiness (lactones) are just some of the aromas issuing forth from the cellulose, hemicellulose, and lignin that make up wood. Speaking of lignin, the other thing that I learned is that woods with higher lignin content literally emit more intense aromas, even before you burn them. As a chemist, I was absolutely delighted to learn about these things. Sometimes, as you go through your professional life, you think you get things mostly figured out. Then, when new information finds its way into your hands, it comes coupled with such a feeling of invigoration.

On that issue of learning new things, I am always amazed by the tenacity at which people who smoke foods at home seek out new information. The number of books and blogs and on-line forums that are dedicated to smoking is just staggering. The intensity of these home-chefs in their search for perfection is matched only by that of home-brewers. As a professional scientist, my life is a constant wading through minutia, checking every dusty corner of published research for that one little bit of information that might transform my research from ho-hum to transcendental. Many home-smokers operate the same way.

Much like research science, a few good results can get you hooked. I started smoking using my Weber kettle grill. Usually reserved for steaks and burgers, I decided one day to try my hand at smoking a pork shoulder. There was something luxurious about having a whole day to look after a hunk of meat. Having a research lab at work and kids at home means that I'm constantly on the go. Carving out a whole day that I get to dictate on my own terms is not a common occurrence. I am more than happy to dedicate a whole day to fussing over my coals and watching my meat slowly cook. While the Weber kettles aren't exactly built for smoking, they can handle a few things really well. And, pork shoulder is one of these. America's Test Kitchen discussed their go-to kettle smoking methods in one of their cookbooks.[7] (Fill a chimney 2/3 full of charcoal and heat the coals until they have been covered in ash. Place several unlit coals, about 1/3 the amount in the chimney, in a pile on one side of the kettle's coal grate. When the other coals are ready, add them to the top of the coals in the grill. Add the food grate to the grill and place the meat on the side opposite to the coals. Put the lid on top of the grill and close down the vents on the top and on the bottom most of the way. After 2 and a half or 3 hours, open the vents all the way. The coals will last for about 4 hours this way before starting over.) Pork shoulder is nice because it's relatively forgiving. Chock-full of collagen, chances are good that you're going to have some tender, juicy meat even if you don't develop a perfect crust on the outside.

Another meat that I like to smoke in the kettle grill is salmon. I set up the charcoal the same way but use a little less (the salmon only takes about 45 minutes to cook). I brine the salmon for a while in a liquid with: water, salt, sugar, garlic powder, and black pepper. Myhrvold discusses the purpose of brining salmon in *Modernist Cuisine*. Brining can add seasoning to the fish and also helps to remove the albumen (the white gunk that seeps out of the fish as you cook it). While I have seen many suggestions for only brining fish for 30 minutes, *Modernist Cuisine* calls for a 24-hour brine to really draw out all of the albumin. This is especially important when cooking fish *sous vide*, but it also helps when smoking. The other trick to smoking salmon that I really love is to place the salmon, skin side down, on a brown paper bag, cut to fit the part of the grill you are using. The paper keeps the salmon from sticking to the grate. Also, the salmon meat pulls right off of the skin, which remains on the paper. It's pretty nifty.

While getting good results is the addiction that keeps you going in scientific research and in smoking, both of these also have the power to humble. I have always loved ribs. It was the dinner of choice for my birthdays when I was growing up. I have always wanted to be able to make ribs for myself. And, slowly but surely, I'm getting close.

The kettle grill, as opposed to a traditional smoker, doesn't have nearly the space required for smoking any reasonable amount of ribs. Traditional smokers look like oil drums, turned on their side, with a little box attached to one end. In these smokers, the smaller chamber holds the wood and the larger chamber holds the food. You control the temperature of the burning wood by opening a vent to the small chamber. You control the convection of heat to the larger chamber by two vents, one between the chambers and another at the end of the main chamber. These vents also help to determine the temperature in the chamber that holds the food. While traditional smokers have heat convection that works horizontally, Kamado-style charcoal smokers have vertical heat convection. One brand's name, Big Green Egg, verbalizes what these cookers look like. The smaller end, pointing down, holds the charcoal. The larger end, pointing up, can hold several layers of food on stacked racks. There are also vertical electric and propane smokers. In all of these smokers, both the heat of the smoke source and the convection through the cooking chamber can be tightly controlled with vents and food placement.

The smoker I have, from the Pit Barrel Cooking Company, is slightly different than these in that I have no fine control over the heat source or the convection through the cooking area. In this smoker, lit charcoals are placed at the bottom of a glorified oil drum, which stands upright. The food is either placed on a grate or hung from racks, both of which are found at the top of the grill.

Of course, the first things that I wanted to try making with this smoker were ribs. I seasoned a few racks of St. Louis Ribs overnight, got some coals ready the next day, and hung the ribs just after lunchtime. They were finished cooking in about 4 hours. And they were perfect. They came out of the smoker with a nice smoke ring, perfectly tender, and with just the right amount of cling between the meat and the bones. I was sold on the idea of cooking like this all the time.

The smoke ring is an odd sight the first time you see it. You expect that if meat is still red after cooking, the color will be found in the middle. A smoke ring shows up on the outside. This curious coloration is caused by the same protein that gives meat its red color, myoglobin. In normal cooking, myoglobin changes from red to brown when the iron atom in the middle of the protein, the one holding on to the oxygen, loses an electron. During smoking, the smoldering charcoal or wood releases a small molecule called nitric oxide, along with all of the other smoky aromas that you are going after. Nitric oxide will displace the oxygen in myoglobin and will permanently keep myoglobin pink, no matter what happens to the rest of the myoglobin structure. While a smoke ring doesn't add much other than aesthetics, it is an indication that your meat was cooked properly. Forming only over the first two or so hours, a good smoke ring only shows up when the temperature in the smoker doesn't get too high at the start of the cooking process. When smoking, you want to balance the temperature so that the meat has time to infuse with smoke, the heat can do its slow work on collagen, the meat forms a nice crust due to the Maillard reaction, and that all of these reach their peaks at the same time. The presence of a smoke ring indicates the gentle cooking conditions that will give optimal results.

The second time I tried to cook ribs in my smoker, my in-laws were visiting. As they were the ones who bought me the smoker, I felt it was incumbent upon me to make them a nice meal with it. Everything was going great until my coals died 3 hours into cooking. I didn't notice until around hour four, when I thought they would be ready, and sadly found that they weren't. Thankfully I salvaged them in the oven, but they weren't nearly as good as those from my first attempt.

The third time I tried to make ribs in my smoker, I got really weird results. One rack cooked up beautifully; they were the best ribs I had ever eaten. The second rack might have been the worst ribs I had ever eaten. Cooked unevenly, they were dry on one end and chewy, from too much collagen, on the other. Granted, I think that this is because I only decided to make the second rack right before cooking, and, perhaps, they didn't completely defrost. But, no matter, I have now made enough "good" ribs that I don't plan on ever being able to stop.

Because I'm a scientist and, now, kind of a smoker, I naturally want to follow obscure information down the rabbit hole. I visit forums to figure out ballpark cooking times and find out what woods and rubs other people use. I also ask as many people as I can. One of the people I trust most on this is my friend Alex Miller, the chemist from the University of North Carolina, who I profiled in the chapter on pancakes. Alex was my first friend to really try tackling smoking at home. With some aspects he's reached a Zen-like state. He told me, "I always used to worry about flames coming from the wood. But, wood is going to burn; you can't avoid it. The most important thing is that the heat moves around the cooking chamber correctly." That was also the conclusion made by a bunch of Harvard engineering students.[8] While taking a class focused on real world problem solving, they tasked themselves with building the perfect smoker, one that would cook meat with as little intervention as possible. They developed and built a smoker, shaped like a distorted hourglass (flat on the bottom, fatter in the middle, more narrow at the top), that supports even convection across all surfaces of the meat that they are trying to cook.

The scientific fascination with collagen doesn't end with roasted meats. Collagen is the cellulose of the animal world. Just as Nature has figured out 1,001 ways to use cellulose in plants, animals are literally built upon collagen. Our bones define our shape and give us strength and support. Our ligaments hold that support together. Our tendons connect our skeleton to our muscles. Our organs provide a home for all of the parts that make our bodies run. This foundation is built entirely of collagen.

In my own research, I try to use the tools that nature has built to do jobs that she never intended. There are many activities in our human world that nature didn't need to come up with solutions for on her own. I am trying to expand the number of roles that natural materials, like collagen, can play in our world, by exposing them to chemistries not found in nature. In one set of experiments I am trying to turn collagen and collagen-like materials into a nouveau-bone constructed with gold instead of calcium. "Why do this?" you might ask. Part of the answer to that question is a rather arrogant, "Because I can." The real answer though, is that our lifestyle is always demanding new technologies and new materials. Using the blueprints that nature has left us, I am accessing chemical tools, designed by humans, to help build new constructs. Part of my science is probing just how far the extremes of this science can go, so that when we are met with technological challenges, we can meet them head-on with confidence.

And collagen is both an inspiration and a tool of the work that I do in my lab. Just like with that pot roast, I expect that a little energy, focused over a lot of time in lab, can transform collagen into new materials that can surprise, delight, and sustain us.

REFERENCES

1. *Ratatouille*, Pixar, Hollywood, 2007, Film, directors Brad Bird and Jan Pnkava.
2. *Beverly Hills Cop*, Hollywood, Don Simpson/Jerry Bruckheimer Productions, The Murphy Company, 1984, Film, director Martin Brest.
3. M. Hazan, *Essentials of Classic Italian Cooking*, Alfred A. Knopf, New York City, 1992.
4. G. A. Di Lullo, S. M. Sweeny, J. Körkkö, L. Ala-Kokko and J. D. San Antonio, Mapping the Ligand-binding Sites and Disease-associated Mutations on the Most Abundant Protein in the Human, Type I Collagen, *J. Biol. Chem.*, 2002, **277**(6), 4223–4231.
5. J. Potter, *Cooking for Geeks*, O'Reilly, Sebastopol, California, 2010.
6. N. Myhrvold, C. Young and M. Bilet, *Modernist Cuisine*, The Cooking Lab, 2011.
7. Cook's Illustrated, *The Cook's Illustrated Cookbook: 2,000 Recipes from 20 Years of America's Most Trusted Cooking Magazine*, Cook's Illustrated, Brookline, Massachusetts, 2011.
8. T. Moynihan, *How 16 Harvard Students Built the Ultimate BBQ Bot*, Wired, 2015.

CHAPTER 12

Color

One of the aims of the modernist cuisine movement is to use all of the tools available (scientific or otherwise) to create the best dining experiences possible. I have highlighted a scant few of these tools and how they are used to prepare food. Many of you will undoubtedly already be familiar with, and know how to use, other techniques that I did not get to.

At the start of any science or technological advance, the practitioners become adept at scavenging the low-hanging fruit, techniques and processes that are easy to grasp and ripe for picking. *Sous vide* cooking, for example, had been described in the late 1700's and actually applied by the food industry in the 1960's. The modernist movement made use of a technique that was there. The same is true for other methods as well. Rotary evaporators, used to extract the essential flavors of fragile foods, and centrifuges, used to clarify juices, are items that are found in nearly any chemistry research laboratory, and have been for the past 50 or so years. Perhaps it is my own laboratory training, but I see modernist cuisine preparations not as haute cuisine, in general, but as ambitiously opportunistic.

In a way, this opportunism has come full circle. Research has begat cuisine. And now cuisine has begat research. As chefs have focused their curiosity on using laboratory techniques in the kitchen, research scientists are now focusing their curiosity on understanding just how these techniques work in a more complicated kitchen environment. In an extreme example of this, several scientists and chefs, including Ferran Adrià, published a study that examined the precise way that calcium ions and alginate polymers interact in the presence of juices to make fruit caviar during the process of spherification.[1]

Not so easily mined are the sensory sciences. How do we perceive what we are eating? What tools and approaches can we use to optimize the dining experience? In laboratory sciences, some researchers are trying to use fMRI, a slight modification to the normal MRI methods that are used to look for cancer cells and torn ligaments and other ailments, to probe what happens in the brain as we eat different foods. While these experiments are really neat, they are also very messy. Eating food while inside of an MRI (a big tube that you sit inside while unseen magnets go whirling past time and again) does not approximate a dining experience. Even determining why different parts of the brain are active during eating (or during any other sensation, for that matter), is a science that is still in its infancy.

The search for underlying scientific truths becomes a little clearer as the studies have moved into the realm of psychology and sociology. Charles Spence and Betina Piqueras-Fiszman have written a book called "The Perfect Meal: The multisensory science of food and dining."[2] Highlighting their own research while discussing the findings of other scientists, their book runs the gamut of the dining experience. How you eat and enjoy your food often has very little to do with what you are actually eating. There are so many design characteristics that are just as important to your meal as the food on your plate. The same food, arranged in different patterns, has been shown to change how much diners enjoy eating it. In these studies, nothing from the kitchen has changed. The seasonings are the same. The meat and vegetables are prepared in the same way from one test to another. They are just plated differently. After eating, the diners rate the meal on a scale. Some food arrangements cause the flavor of the food to be rated higher than others.

Nearly every aspect of the dining experience, from the design of the restaurant's exterior, to the lighting and sound, to the heft of the cutlery, to the presentation of your bill, has been studied. All of these little touches grab hold of your unconscious, which is always weighing how much you are enjoying yourselves. That many fast food restaurants use red and yellow signs and logos is a reflection of the fact that red induces feelings of hunger and yellow is the color that most quickly catches our attention. It should also come as no surprise that restaurant managers really care about the sound properties (music and ambient noise from conversation and the kitchen) of their dining rooms. Read any critic's review or on-line discussion and the sound level and ambiance is always described. More extreme forms of studying sound and food are examples where diners are put in a state of sensory deprivation, where their only experience is that of eating; they eat in the dark while wearing noise-cancelling headphones.

How the findings from these studies get applied though, is another story. As opposed to the technological aspects developed in laboratory research, applying the understanding from psychological studies can be much trickier. Just how do you find balance in your plating? What type of music is best to play during dining? A lot of this may seem like an exercise in trying to control diners. That is, "How can we make people like their food, and be willing to pay for it time and again, by tricking them with lights and sound and organization?" We humans may be easily led at times, but I don't think that is the real goal of these studies. What I see the dining experience as, is as an art. The more advanced you move into any art, the fewer people there will truly be to recognize and appreciate your work, much less be able to truly pay for it. Restaurateurs must always balance their desire to push the limits with the expectations of their customers. Some restaurants and meals are sought out because the diners know that they are going to be getting an eye-opening and surprising experience. Certainly not all dining can be thought of this way.

But that does not negate the fact that our psychological makeup, along with our chemistry and biology, affects the way we experience food.

Of all of the senses involved in eating, sight is certainly the primary. Not only does light move faster than sound can, or than aroma can waft, but our brain will often process visual signals first. The fact that McDonalds can make us feel hungry now just by using red and yellow signage, is one way that sight affects our eating. But color and contrast can also affect how food tastes.

The most shocking, and subsequently well-told, example of this is from a study of the perceptions of red and white wines.[3] Some students, who were studying winemaking, became the unwitting subjects of an experiment on how visual cues change the way we taste. Before tasting several wines, the students agreed on typical flavor descriptions that could be used for either red or white wines. While red wines are usually warm and can be plummy with hints of tannins, white wines are crisp and citrusy and more floral. The students then tasted two wines, a white wine and a red wine. The trick in this study is that the red wine was actually just the same white wine with added food coloring. The students described the white wine using the terms typically reserved for white wines, and they described the red wine using the terms typically reserved for red wines. This experiment does not prove that the students had no clue as to what they were doing. What it shows is how much of what we perceive as flavor, can be defined by senses that aren't remotely connected with taste buds or olfaction.

Returning to the theme of fine dining as art and the importance of plating, there are many chefs who don't just plate their food artistically; they plate their food to resemble art. Whether you are trying to be creative with your plating or not, I think that most people would agree that any dish benefits from a little color. In the palate that we have for cooking, all of our vibrant colors come from vegetables.

In the process of researching this book, I had a conversation with my friend, Gretchen Keller, with whom I worked at Caltech. Gretchen has a PhD in chemistry and also happens to be an artist. When we were talking about color, she said, more eloquently than I could, "Most of the food that we eat is pretty drab. Meat and grains end up brown. All of the vibrancy and contrast and life come from vegetables."

My own scientific training has, in many ways revolved around color. As an undergraduate, I made bucky-balls and nanotubes working with Howard Knachle at the University of Dayton. Bucky-balls (also called buckminsterfullerene or C_{60}) take their name from Buckminster Fuller who designed architecture that resembled these molecules.[4] But, C_{60}, which won its discoverers the Nobel Prize in 1996, better resembles a football (a soccer ball to my American friends). A series of carbon atoms arranged in hexagonal and pentagonal shapes, wrapping around in a circle. Working with them in lab, I was absolutely taken by their beautiful purple color. Add an extra 10 atoms (going from C_{60} to C_{70}) and they look unmistakably like red wine.

But my understanding and appreciation of color didn't take shape until I went to work for Harry Gray at the California Institute of Technology as a postdoctoral researcher. Harry is one of the guiding lights in the field of inorganic chemistry and is specifically credited with building the modern sub-discipline of bioinorganic chemistry, which is the study of how metals interact with biological molecules and how that interaction changes both the chemistry of the metal and of the biological molecule it associates with.

Harry, besides being a brilliant chemist and warm human being, is a fantastic storyteller; he has one for every occasion. Perhaps his greatest story is the retelling of how he started his scientific career.[5] As a young teenager, Harry found himself fascinated by colors. The color of plants and leaves and paints and clothing were a mystery to him, and he needed to know more. One thing led to another and he figured out that he could isolate the dyes from clothing if he could get his hands on some concentrated sulfuric acid. So, he wrote to a chemical supply company, claiming he was a professor at the local university, and requested some sulfuric acid for his research. The chemical company replied in kind and sent a bottle to Harry's house. Emboldened by his unlikely success, Harry decided to tempt fate again and actually try to isolate some dye molecules. Knowing Harry to be the brilliant chemist that he is, it is no surprise that he did end up performing a successful experiment. Unfortunately for Harry, he had decided to use his mother's clothing. After she found him out, I'm not sure which was redder: the dye Harry had extracted, his mothers face, or Harry's backside.

Harry was hooked. He has spent the better part of his career trying to understand color and what causes it. Before he became a faculty member, he moved to Denmark in the 1960's to help develop a new theory for describing the colors of metals and their compounds. Even though he hasn't directly worked on color chemistry for quite a few years, his interest in this topic hasn't lost any of its vigor. During my time at Caltech, whenever anyone would make a new metal complex, the first thing Harry would ask is, "What color is it?" He usually had a guess as to the color and he was usually right.

His interest in color, it turns out, has more relevance to chemistry than just its aesthetic properties. The color of a chemical is often directly related to its reactivity and ability to change into something new, or foster the transformation of other molecules.

Chemistry is governed by electrons. When we learn chemistry, usually we are introduced to the periodic table. All of the different elements lined up in order from start to finish by the number of protons that are found in their atomic nucleus. The shape of the periodic table, the reason why it is tall on either end, the reason why it dips in the middle, the reason why there is that extra little appendage running parallel to its bottom, and the reason why they are organized in columns, is because of their electrons. Electrons, which are 2000 times lighter than protons, define the reactivity pattern of all of the elements. For instance, lithium and sodium and potassium are all in the same column because they react the same way with chlorine.

The ability of electrons to control chemistry is not limited to individual atoms. Molecules are built with electrons as the mortar; the electrons act as the glue that holds atoms together in bonds. And, when a molecule goes through a reaction to gain a new chemical identity, the electrons in the molecule are what dictates how that transition occurs. The energy and location of an electron will dictate whether it will take part in a reaction or not. And, by extension, these electron properties dictate the reactivity of its host chemical.

So, what does any of this have to do with color and what does it have to do with vegetables? It turns out that color, like reactivity, is dictated by the energy of an electron and its location within a chemical.

Thinking about the light that we are able to see, the visible spectrum (the red, orange, yellow, green, blue, indigo, and violet), each color in the rainbow has its own unique energy. Red has a lower energy than green, which has a lower energy than blue. Each of these colors will affect electrons in different ways.

Electrons are subject to the rules of quantum mechanics, which state that only the perfect amount of energy can get an electron to change its energy and location. The higher an electron's energy, the more likely it is to move and be involved in a reaction. However, the rules of quantum mechanics are different than what we are used to in our daily lives.

My kids used to love being pushed around in a shopping cart at the grocery store. When I was in charge, I would find an open aisle and take off at a sprint. They would get such a kick out of that. If there were obstacles in my way, like display cases or people, I could slow down accordingly. The point of this story is that I could push them at any speed I liked. If my children were electrons and I was trying to excite them (get them to move from a low energy to a higher energy and make them reactive), I would not be able to push them however I wanted. If I pushed them too slow, they wouldn't move. If I pushed them too fast, they wouldn't move. I would only be able to push them at one of several prescribed speeds, with corresponding energies.

The energy that electrons need, to go from a low energy to a higher, more reactive energy, turns out to be similar to the energy contained by the hues of visible light. With the spectrum of colors in the visible light there is a spectrum of energies. An individual electron within a molecule might only be pushed by a single part of that spectrum. But a molecule contains many electrons and so, as a whole, that molecule's different electrons can be pushed by different colors of light.

Perhaps an example is in order here. Chlorophyll is the chemical that is responsible for giving plant leaves their green color. Chlorophyll contains a central magnesium ion surrounded by a complex ring of nitrogen, carbon, oxygen, and hydrogen atoms, with one part of that ring sporting a little tail. The electrons in chlorophyll are pushed by very specific colors (energies). Purple light and a schmear of yellow and orange and red, are effective at pushing chlorophyll's electrons around, with blue light being slightly less effective. If we send white light (light that contains the whole visible spectrum) through chlorophyll, the ability of different colors to push electrons from a low energy to a higher energy, is noted by the amount of each color that is able to pass through chlorophyll. For chlorophyll, because purple, blue, yellow, orange, and red, get used to push its electrons, green is the only color that is able to pass through. This is the reason why chlorophyll, and the leaves that contain it, are green. When sunlight hits a leaf, most of the light gets used to alter the energy of chlorophyll's electrons. The rest of the light, the green part of the spectrum, the stuff that doesn't get absorbed, bounces off of the leaf where it can potentially meet our eyes.

COLOR

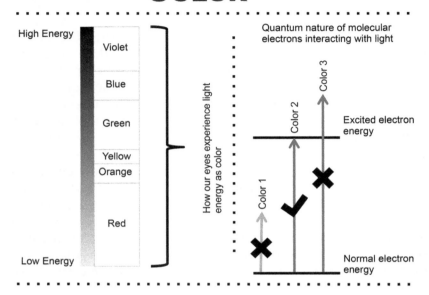

Chlorophyll: the green color of our leaves

chlorophyll a

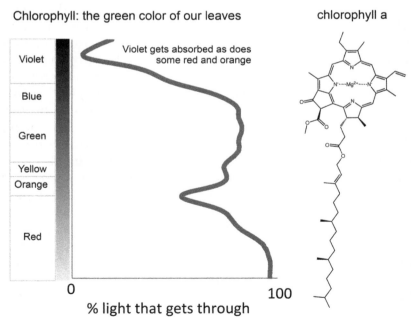

Violet gets absorbed as does some red and orange

% light that gets through

Color

The way that chlorophyll interacts with light is doubly appropriate for the content of this chapter. The first is obvious, the green color of leaves in our food is a burst of color that makes our lives interesting. The second involves the use of chlorophyll in photosynthesis, combining two of Harry's great contributions to science over his career: explaining the color of metal compounds and understanding the way electrons move from one place to another. In photosynthesis, the process by which plants use sunlight to convert carbon dioxide and water into sugar and oxygen, the chlorophyll is the part of the plant that captures the sunlight, which excites chlorophylls electrons to a higher energy. One of these high-energy electrons has enough energy to break free and run away from chlorophyll so that it can be used in the reaction that eventually turns carbon dioxide and water into glucose and oxygen.

My view of color is necessarily linked to Harry. His scientific achievements and enthusiasm for color are ingrained in me as they are with all of the people who have worked with him over the years.

Because everyone perceives color differently, I decided to talk with several of Harry's former graduate students to understand how they view light and color and how they use it in their cooking. Jillian Dempsey, Morgan Cable, and Gretchen Keller all began their PhD training at the same time that I joined Harry's lab as a postdoctoral researcher. One of my roles in the lab was to help oversee the work of these burgeoning scientists. I have always felt close to them because of this connection. They are all, in their own unique ways, absolutely wonderful people whom I am glad to have as friends. They're answers to my questions about color all have similar shades; we all have the same training. But, each of the responses gives a little insight into these young scientists, their own lives, and their own work.

JILLIAN DEMPSEY

Jillian Dempsey is a professor of chemistry at the University of North Carolina in Chapel Hill. While she was at Caltech, Jill was part of a team looking into new ways to store solar energy. More specifically, Jill studied a set of reactions that turned protons (H^+ ions) into hydrogen gas (H_2) that could be used later to generate electricity. While most solar panels only provide energy during the daytime, Jill's research worked on understanding how to trap that energy in chemical bonds so that we have access to clean energy when the sun goes down. Jill has expanded on this work since moving to UNC. While the goal of her research is to make efficient solar energy devices, the day-to-day work in her lab focuses on understanding all of the little chemical steps that will make this possible. Because we can't see individual molecules and accordingly can't watch chemical reactions with our eyes, Jill's team does something really ingenious. They make molecules that change color at each one of the chemical steps in the transformations involved in solar energy storage. So, even though they can't watch the atoms, the colors of their molecules and the speed at which they change from one color to another, tell them a lot about how the reactions are working.

I asked Jill what colors mean to her and what emotional response she has to them. The first thing she said, without hesitation, is that when she sees bright and vibrant colors, she knows that things are going to go well in lab. This is in contrast to when her reactions produce brown or black colors, which mean just the opposite.

Jill told me about one of the first reactions that she ever ran in a research lab. In the first step of the reaction, she mixed two chemicals together, which resulted in a brilliant magenta color. To finish her task though, she had to add a third chemical. She nearly refused to do that however, because she knew that the magenta color would go away with the second addition. And just like the changing colors in a chemical reaction, Jill said that her favorite time of year is the spring, when greens bursts from drab browns.

In every chemistry "show" or demonstration that I have ever watched, the showstoppers are reactions that make color or reactions that make fire. If the reactions make colored fire, well, then, look out! Physicists have stars. Biologists have animals. Chemists inspire awe with color.

That inspiration doesn't stop in the lab, either. Jill is a phenomenal cook. My favorite times with her have always been when our families have shared meals. Jill sent me one of her favorite recipes from Yotam Ottolenghi's cookbook "Jerusalem: A Cookbook."[6] Because I know that Jill adores his cookbooks, and because my

Color 223

wife and I generally adore Jill's cooking, I asked her why she's drawn to his food. She said that he combines ingredients in unconventional ways and that brings out unexpected flavors. It is his creativity in doing this that she's most drawn to. The recipe that she shared is for roasted butternut squash and red onion with tahini and za'atar. (Roast 1 squash, cut into pieces, with 2 sliced red onions at 240 °C (475 °F) for 30 or 40 minutes. Make a sauce by combining 53 mL (3 1/2 tablespoons) of tahini paste, 23 mL (1 1/2 tablespoons) of lemon juice, 30 mL (2 tablespoons) of water, 1 crushed clove of garlic, and a pinch of salt. Toast 53 mL (3 1/2 tablespoons) of pine nuts. Plate the squash and onions. Top with the sauce, za'atar (a blend of Middle Eastern herbs), and fresh parsley.)

The contrast between the orange of the squash, the maroon of the onions, the white of the sauce, and the green of parsley results in a warm and inviting plate of food. The carotenoids from the squash are exceptionally striking in this, having been given an extra jolt of gold around the edges while roasting. These are interesting molecules. Obviously named because they

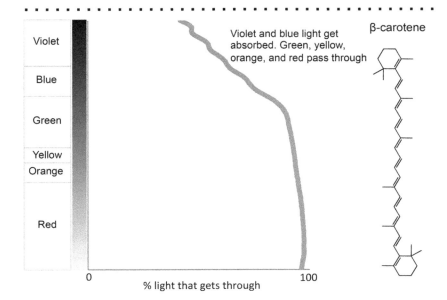

SQUASH: CAROTENOIDS

are also found in carrots, carotenoids are long molecules that get their color from the alternating bonding pattern along their length. They absorb violet and blue light, allowing the green, yellow, orange, and red light to pass through or reflect off of the squash. When we look at squash, we don't see green, yellow, orange, and red all at once. We see an average of these, which turns out to be a yellow-orange hue. Carotenoids are especially interesting in nature in that they provide a natural sunscreen for chlorophyll, blocking the damaging violet light and allowing the red light, the light that moves chlorophyll's electrons, to pass through.

MORGAN CABLE

Morgan Cable has more energy than any human I have ever met. She is entertaining and inspiring and exhausting all at once. While she was at Caltech, Morgan worked on ways of detecting anthrax spores. The reasons for this were twofold. First, it would be good if we were able to easily figure out if there were anthrax floating around. The second reason was for a completely different application. Morgan worked jointly at NASA's Jet Propulsion Laboratory in nearby La Cañada. She was interested in anthrax because it can survive as spores, some of the toughest forms of life on Earth. Before NASA sends any probe into space, it needs to make sure it is free from any potential biological contamination; we don't want to be responsible for colonizing Mars with bacteria. Morgan developed a way to search for anthrax and other spore-forming microbes by exploiting color. Specifically, she developed chemicals that emitted light in the presence of anthrax. She got to work with some really interesting metals while doing this research too: terbium (glows green), europium (red), dysprosium (yellow), and samarium (pink). And Morgan got to travel all over the world in search of bacteria that live in harsh environments, including the Atacama Desert in Chile, one of the driest places on Earth.

Morgan is still at JPL, where she is a research scientist, among other duties. She is still working on probes being sent to space. But now, she helps to develop landers and figure out what types of chemical measurements they should be making when they reach their destinations. When I was talking with her, she was just about to take off to a meeting discussing the Europa lander. (Europa is one of the moons of Jupiter. It has a subsurface ocean, which many other moons do, but its ocean is closer to the surface than any of the others.) Morgan's work, in effect, has gone from searching for resilient life on Earth to searching for resilient life in far away worlds. Morgan is also trying to create an environment in her laboratory that mimics these far away worlds. One of her current projects involves setting up an experiment to figure out why the beaches on Titan, the largest moon of Saturn, have bathtub rings. This astonishes me to no end. That we know that Titan has soap scum-like rings around its lakes is one thing. Setting up an experiment to try to recreate them is an exercise in amazing.

Aside from her incredible work, everyone who knows Morgan expects that she will become an astronaut some day. Whenever my kids ask about NASA or the space program, I say with absolute sincerity and certainty, "I know someone who is going to become an astronaut!" Morgan doesn't have an "in" or anything like that. It is one of her life's goals, and I know it will happen. So, you saw it here first. Someday you will be reading a news story about my friend Morgan, the astronaut.

When I asked Morgan about color, ever the analyst, she said that color is how we interpret the world. The full range of light is vast and the part that we can actually see is so very small. She told me about mantis shrimp and their amazing capacity to see light.[7] Cone cells are the specialized cells in eyes that allow animals to see color. We have three types of cone cells that allow us to see individual colors or a blend of colors from the visible spectrum. Dogs only have two kinds of cone cells. But, the world of light is so much more interesting than what we see. Bees can see ultraviolet light (which has a higher energy than visible light). To a bee, flowers look markedly different than what they do to us. The mantis shrimp has an incredible number of different cone cells; 16 to be exact! That's a lot of color for one little animal to be able to see.

Morgan also told me that she uses color to organize her thoughts and her schedule. She has been doing that since high school. For her, each color has a different meaning: blue for headers and red for something important. She said that her office is a color-coded panoply of tasks and thoughts.

WINE: ANTHOCYANIDS

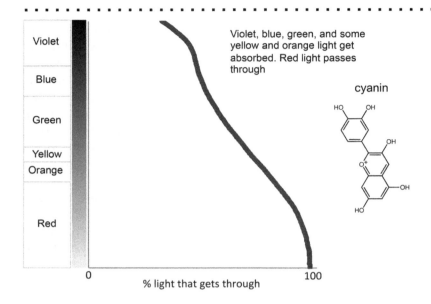

Morgan sent along her favorite colorful recipe: purple chicken. It's really just a modified *coq au vin* recipe, but I think I might like Morgan's name for it a little better. (In a pan, brown some chicken thighs, bone-in and skin on, in olive oil. Remove to a plate. Sauté onions and a little garlic in the same pan. Replace the chicken, skin side up, and fill the pan with red wine. Cook slowly on the stovetop or in a pre-heated oven.) As it cooks, the chicken meat infuses with the color of the wine. On its own, the chicken would be a bland brown, but it takes on a nice purple color while braising, hence "purple chicken". Wine color is really interesting. Some colors arise from individual molecules; cyanin absorbs blue, green, and yellow light. The result is a wine colored wine (naturally). The color from these molecules is used by nature to attract animals to eat the fruit and, eventually, scatter its seeds. Other colors develop through wine aging, when cyanin and similar anthocyanidins react with molecules produced by the yeast during fermentation; these modified chemicals result in modified colors.

GRETCHEN KELLER

Gretchen Keller came to Caltech with a force of personality that I don't think any of us were really ready for. Chemists (and most scientists, really) are perceived to be quiet and shy and maybe a little aloof. While this doesn't actually reflect reality, there is some bit of truth in that observation. Gretchen came to grad school already a bright chemist, an artist, and an organizer of people. People are just naturally drawn to her. At Caltech, Gretchen worked directly with me, which is probably why nothing went right for her in lab until after I moved to Maryland to start my job as a professor. Amazing how that works.

Gretchen studied the way proteins take or lose their shape. I have talked about protein unfolding a lot in this book. But Gretchen studied it from a different perspective. Gretchen's work focused on the protein that is responsible for plaque formation in people suffering from Alzheimer's disease. As her stepmother suffered from and succumbed to a similar aging-related disease during Gretchen's PhD work, her experiments took on an extra layer of meaning. I can't stress how difficult this protein is to work with. If you put it into water, it starts to aggregate, just like it does when making plaques in the brain. And, like every bit of research in Harry's lab, Gretchen used color to probe how this Alzheimer's protein wibbled and wobbled its way through water.

Gretchen is still a woman with many talents and a big heart. She works for an organization that ensures the well-being of laboratory research animals. In her spare time she teaches art classes. I was excited to talk to Gretchen because I knew her thoughts on color and composition have been formed differently than Jill and Morgan's, let alone my own.

Gretchen does look at colors from two points of view. When she is painting, she can't help but be the curious chemist and wonder about the chemicals that give rise to the colors in the pigments. (This sentiment was shared by Jill who loves the fact that, after graduate school she knows how to identify the chemicals in paintings.) For, Gretchen, color also corresponds to emotion. When Gretchen paints, she starts in blacks and whites and sepias and abstractions. She told me that most of her own paintings stay in these abstractions; she likes the contrast that these muted tones provide to the shapes that she paints. Most of these paintings are landscapes. When she paints people though, she adds colors that reflect her own emotional ties to the subject. Gretchen made sure to impress upon me that adding color means layering and blending and contrasting many colors to create real vibrancy. Emotions are never as simple as single thoughts or single feelings or single colors. Life is messy and beautiful all at the same time.

When it comes to food, Gretchen says that fruits and vegetables are what bring that vitality and life and those meals that don't include these are necessarily dull. Because fruits and vegetables provide the colors of our meals, colors come to mean life as well. In parallel with her painting, she seeks out foods that blend and contrast many colors.

The recipe that she sent me, a breakfast tofu scramble, is no different. Drain 225 grams (8 ounces) of firm tofu. Sauté your choice of vegetables in a pan, making sure to choose some greens and reds. Crumble the tofu and coat it with at water based slurry containing: a pinch of salt, garlic powder, cumin, chili powder, and turmeric. In the recipe she sent, the turmeric gives the tofu an egg-like appearance. But, what's important about the turmeric is the way it ties the other colors (the reds and greens of the vegetables) together.

Turmeric gets its intense yellow color from a molecule called curcumin. Working with turmeric seasoning is not for the faint of heart. Should you decide to cook with it, you need to commit yourself, knowing that you will stain your hands or your clothes or both. This warning is even more relevant when working fresh turmeric. Curcumin absorbs only violet light. The light that does not get absorbed can be detected by our eyes and is seen as a powerful yellow. Much like the carotenoids, there is a definite pattern to the bonds within the molecule. But, curcumin is not nearly as long of a molecule as any of the carotenoids. That length turns out to be really important for defining curcumin and carotenoid color. The longer one of these molecules grows, the more and more it will absorb into the blue then green then yellow, changing its color from yellow to orange to red.

ME

So, what do I think about color? Well, like my former co-workers, I can't help but be swayed by my training. I love thinking about the vibrancy that pigment molecules bring to art and to food. I

Color 229

TUMERIC: CURCUMIN

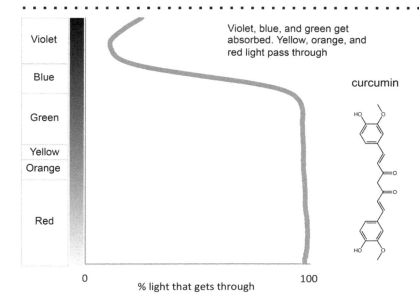

too look at paintings and wonder what chemistry makes their colors possible.

I want to share my own recipe before I add any more. Beets are not one of my favorite foods. Erika loves them. I suppose that I had only ever seen them in canned form. (Erika loves these, too.) But, I love the way that this recipe uses a little bit of citrus to make them come alive and how the creaminess of some feta brings it all together. But, most importantly, how can you deny the beautiful red and orange and purple of roasted beets? (Roast 2 or 3 large beets with olive oil and salt at 204 °C (400 °F) until they are tender. Zest an orange and reserve. Juice half of the orange and use the juice to make a vinaigrette. I don't find orange juice to be acidic enough for my tastes here. I add a little pinch of citric acid to embolden it. After the beets are roasted and cooled, peel them and dice them into cubes. Dress them with the vinaigrette and crack fresh pepper over them. Top with some crumbled feta or goat cheese and mint leaf.)

The best part of eating beets is how their pigment finds its way into your urine. Nothing like being surprised with red urine,

BEETS: BETANIN

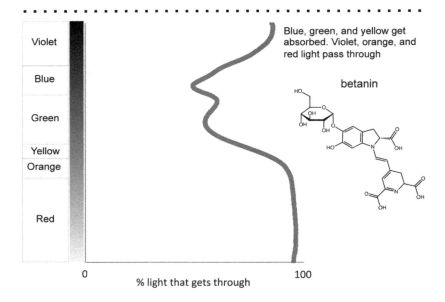

having forgotten that you had eaten some beets! This red color comes from a group of molecules called betalains. The most commonly studied of these is betanin. And, obviously, we don't digest these molecules. They pass out of us looking exactly the same as when we ate them. In the plant world, these pigments are fairly rare, which is probably the reason why we can't digest them. Betanin absorbs all light except for red. Scientists really aren't sure how these molecules are used in nature. But, I'm glad that she does use them. They provide so much vitality to our world of cuisine.

Now to how I view colors.

Colors are deceptive. They show the world everything that an object is not. The color that bounces off of a material is everything that the material doesn't keep for itself. In that way, colors are a lie, a projection of what something was unable to be. But there is beauty in this too. The lie of projection allows us to see clearly the truth and chemistry of what an object holds.

At the start of this chapter, I discussed how organization of food on a plate can "trick" diners from liking the food more.

Colors trick us as well. But when we fill our plates with contrasting colors and contrasting projected lies, the held reality of one ingredient is balanced by the held reality of another ingredient. Lies give way to truth, our reality from the color in the foods on our plate.

REFERENCES

1. H. Fu, Y. Liu, F. Adrià, X. shao, W. Cai and C. Chipot, From Material Science to Avant-Garde Cuisine. The Art of Shaping Liquids into Spheres, *J. Phys. Chem. B*, 2014, **118**(40), 11747–11756.
2. C. Spence and B. Piqueras-Fiszman, *The Perfect Meal: The Multisensory Science of Food and Dining*, John Wiley & Sons, Oxford, 2014.
3. G. Morot, F. Brochet and D. Dubourdieu, The Color of Odors, *Brain Lang.*, 2001, **79**(2), 309–320.
4. The Royal Swedish Academy of Sciences, *Press Release: The Nobel Prize in Chemistry 1996*, [document on the Internet]. NobelPrize.org; 1996 October 9 [cited 2016 July 5], available from: http://www.nobelprize.org/nobel_prizes/chemistry/laureates/1996/press.html.
5. H. B. Gray, *Voices of Inorganic Chemistry – Harry B. Gray*, [video file]. 2011 August 1 [cited 2016 July 5], available from: https://www.youtube.com/watch?v=Si7bjzkD-Fs.
6. Y. Ottolenghi and S. Tamimi, *Jerusalem: A Cookbook*, Ten Speed Press, Berkeley, 2012.
7. R. Krulwich, *Don't Go Near the World's Champion Rainbow Watcher. It's Mean. Very Mean*, [document on the Internet]. National Public Radio; 2013 April 10, available from: http://www.npr.org/sections/krulwich/2013/04/10/176807785/dont-go-near-the-worlds-champion-rainbow-watcher-its-mean-very-mean.

Drinks and Dessert

CHAPTER 13

Beer

The first time I ordered a beer in front of my dad I was 19 years old. This probably doesn't seem like that big of a deal for people outside of the United States, where the legal drinking age is under 21 years old. But, I decided to tempt fate and see how he would react. My parents weren't overly strict when I was growing up; they did have very high expectations for my brother and sister and me, and we were held to those. Mostly, though, my parents had a strong sense of moral certitude. (Now that I'm a parent and I exercise my own "moral certitude" on my children, I understand how conviction can often be a show to raise them right while helping to keep your own sanity.) On the occasion of my alcoholic apostasy, we were on vacation, and I thought, "Why not?" After I placed the order, I vividly recall the very stern look that my dad gave me. He looked at our waiter and said, "He can't do that." I looked back at him and said, "Yes, I can." Dad looked like he was going to have it in for me. (I still laugh thinking about this.) "Dad," I told him, "We're in Canada. I'm totally allowed." My dad had forgotten the drinking age in the Great White North. I was a 19 year old who lived a 4 or 5 hour drive away from the Canadian border; I had obsessed over the legal drinking age in Canada for a while. There was a smug satisfaction on my face as I drank my

---

Chemistry in Your Kitchen
By Matthew Hartings
© Matthew Hartings, 2017
Published by the Royal Society of Chemistry, www.rsc.org

beer, catching looks from my parents during dinner. That Labatt's might have been the most satisfying beer I've ever had.

I know that my dad has his own nostalgic relationship with beer, too. He travelled to Europe (mostly Germany and Austria) when he was younger. I really don't know how much of the local beer culture he and his friends partook in while they were there; they were travelling with Catholic nuns. What I do know is that when we travelled to Munich for a vacation just after I graduated from college, the first thing we did upon picking up our bags and our car was to head directly to the Hofbräuhaus. His excitement was palpable. We had been on the plane all night and were exhausted. Well, most of us were exhausted. Dad was giddy. We all enjoyed various sized steins of Munich's finest before touring the rest of the city. From Munich, we drove to Andechs and stayed at a guesthouse not so far away from the Andechs Abbey where monks oversee a working brewery. Through the haze of jetlag and probably too much beer, all of us really enjoyed Andechs.

On a recent trip to visit my in-laws, I was surprised to find several varieties of Kloster Andechs beer at a store in Boardman, which is a little town that sits in northeast Ohio. Who would have expected beer from an independent and somewhat local brewery to turn up in a small town in Ohio? I bought some for my dad. I think he was pretty thrilled to have it again. And after drinking it, he started hounding his local beer distributor to see if they could find any more for him. So, yes, the beer nostalgia does run strong in my family.

My parents never really had very much alcohol in the house while we were growing up. Beer (and maybe wine) were saved for having company over to the house or for throwing a party. Sometimes my dad and grandfather would have a beer during a break on one of our many household projects, which come in a fairly constant stream when you live in a 100-year old farmhouse. Despite the scarcity of beer in the house, or perhaps partially because of it, sharing a drink with my father always marks an occasion.

When Erika and I were living in Chicago, a drive home only took about 5 hours if we were lucky to avoid traffic. So, we tried to make it back to Ohio whenever we could. Without fail, my dad would offer us a beer as soon as we walked in the door, always ready to celebrate. On one particular trip back, we couldn't leave from Chicago until later in the evening. We didn't make it to my childhood home until well after 1 in the morning and probably much closer to 2. Dad was still waiting up for us, lying on the couch and watching television. This in itself was not unusual. When my siblings and I were in high school, and later when we were in college, he would always wait up whenever we went out with friends. The difference on this particular drive of ours from Chicago is that when we got home and woke Dad up from his TV watching at 2 in the morning, he was ready to have a beer with us. After a day of putting in overtime at work and driving 5 hours, a beer was not the first thing on my mind. But, Dad was ready to celebrate and so were we.

BREWING: BIOLOGY AND CHEMISTRY AS ART

Being a chemist, I had always wanted to brew my own beer. We lived in fairly cramped quarters in California; Erika and I were in one bedroom and our two girls were in the second. Our living space was filled with baby and toddler toys. It was tight. There was no way that I was going to be able to brew a bunch of beer in that apartment.

When we moved to Maryland, I finally had the space to store all of the necessary beer-making equipment and supplies. For my first attempt, I decided to try making an ESB.

I was first introduced to the quality you get from homebrewing by the husband of one of Erika's former colleagues. When Erika was pregnant with our oldest, she had a baby shower at work. As one of "our" gifts, we received a set of growlers from our homebrewing friend. The best of the bunch was a delightful ESB, still one of the tastiest ESB's I've ever had. Importantly though, it showed me that homebrew could stack up to the best craft brewing has to offer. I had been under the impression that making beer at home could just be a fun thing to do and would be a way to have something interesting to share with friends who came for a visit. But, his beer was balanced, rich, and flavorful. Admittedly ambitious, this is what I wanted to recreate with my first dive into the world of homebrewing. If nothing else, I could learn a lot from striving to reach this perfection.

So, I ordered a kit from a beer making supply company. The options are really overwhelming and range the gamut from as easy as humanly possible to seemingly impossibly complex. On the simpler side, there are companies like Mr Beer. On the more complex side, you can create your own recipes with custom blends of grains, custom blends of hops and other flavorants, and custom blends of yeast. I decided to take the middle road. "I am a chemist, dammit," I thought. "I can handle a little complexity." The kit that I ordered was somewhere in the middle of the complexity spectrum, probably skewing a little closer to easy than this chemist's arrogance finds easy to admit. It came with some malt syrup, some malted grains, a clarifying agent, some hop pellets, and some yeast.

The equipment that I purchased included a bucket and a large glass jug (for different stages in the beer fermentation), some tubes for transferring liquids, and other various accessories for cleaning, sealing, and storing. What I didn't buy was a large pot for prepping the liquid that would eventually be turned into beer. We had a stockpot at home, and I thought that we would be able to make good use out of that.

The kit that I purchased was for a 5-gallon batch of beer. My stockpot, I found out, only holds about 3 gallons. As a chemist, I knew that my ability to maximize the amount of sugars and flavors that I could extract from the grains was directly related to the amount of water that I used. So, I added as much water as would comfortably fit in the pot. When it came time to boil this liquid and add hops, the pot was pretty full. I got it going to a nice gentle boil and thought I was in good shape. Well, that was until the gentle boil turned into a rolling boil and about a half gallon of that sugary liquid boiled over the pot and ended up all over my hot stove. If you've never cleaned up burnt syrup from a stovetop, count yourself as lucky. (It took me about 3 hours and lots of scrubbing to get the stove back in working form after this little mishap.)

I was a little panicky at this point about how the beer would turn out. But, as Charlie Papazian says, "Relax. Don't worry. And have a homebrew."[1] Papazian is an engineer, by training, and is probably the most important influence on the development and growth of American homebrewing. His book, "The Complete Joy of Homebrewing," is required reading for anyone who is looking to make beer at home. Knowing how intimidating the different parts of brewing might seem to some, he repeats his, "Relax," mantra a number of times throughout the book. What he's really trying to say with this is, "So what if you screw up? Your beer is still going to taste good when you've finished." And for the most part, that's really true. Even the big spill that I had didn't ruin the beer I was making, which was delicious, by the way.

Much like roasting meat, I find brewing beer to be a bit of a luxury and rather cathartic. Time is always the most difficult thing to find in any day. And having the time to brew beer is a wonderful feeling in and of itself. I have made a number of different beers since my first attempt. I learned a little more about brewing each time I do so. (The one constant in all of this is making sure you have properly cleaned and sterilized all of your containers; wild yeast or bacteria can really send your beer off the rails.) And Papazian is right. No matter how intricate and precise you get in your beer making, it is going to taste pretty darn good when you are finished.

Brewing beer highlights the difference between research chemistry and kitchen chemistry in a very extreme way. Research chemistry usually involves an attempt to understand and optimize an individual reaction while cooking chemistry aims to find a way to have many different reactions finish at the same time. Brewing is certainly an example of this, having various food chemistry reactions conclude at once, taken to an extreme. Each batch of homebrew that I make takes about a month from start to finish. Along the way, there are several critically important chemical reactions that happen that are the focus of the brewer. But, while these are going on, there are other chemistries occurring that are no less important and which give each batch of beer a bit of vibrancy and uniqueness.

The more I brew, the more I compare it to parenting. There is a core set of values that you want to instill in your children. (Every style of beer has an alcohol content, appearance, and flavor profile associated with it.) You define the "moral certitude" that you think will give your children the core that you want them to have. (Beer ingredients are chosen that should result in the style that you want to produce.) You expose your children to life lessons where you can apply, and they can see and understand, your convictions. (You carefully choose each step in the brewing process so that your target beer style will be produced in the end.) You send them off on their own to interact with others and be changed, expecting that the core that you have given them will stay intact. (You add yeast cells to your developing beer with the knowledge that it will be changed over time and expecting that your target qualities will be enhanced by the way the yeast changes your beer.) From the beginning, the way that you choose to implement your parenting will necessarily have unintended consequences that also influence the way they grow and mature. (Your ingredients and the way you choose to process them come with complexities, which although they are miniscule on their own, collectively influence the beer that you produce.) Sometimes we understand what these unintended consequences will be; sometimes we don't. Either way, your children grow and develop personalities of their own. That, I think, is the best part of being a parent. The beer that you brew at home will also develop its own character, enhancing and deepening the style you had originally intended to craft. And, as with many other foods that we prepare, the careful dosing of extra time can do wonders to bring out new personality to beer.

While alcohol content, appearance, and flavor profile are the ultimate traits for your beer, malted grains, hops, and yeast, are the ingredients that will bring you there. The brewing scene in the States has been dominated by talk of hops in recent years. However, the most interesting part to me, and the most interesting chemistry in brewing comes from the malt.

ALL ABOUT THAT BASE

Barley is the backbone of any beer recipe and the base upon which that beer is built. The type of beer that gets brewed is entirely dependent upon how the barely gets modified after it is harvested. That this is the case for barley is due, in part, to some nifty chemistry and, in part, a fortunate coincidence of nature. Barley is the grain of choice for two main reasons. Each stalk produces a lot of starch, starch that can be converted into sugars, which will eventually be turned into alcohol through fermentation. Barley also contains the requisite proteins that can turn the starch into fermentable sugars. These proteins and starches are the basis for both beer and whiskey. When we discuss beer making, we really only talk about the protein and the starch. But, the barley that we use for our libations is not just some simple mixture of these two chemicals.

Like all plants, barley, which is a type of grass, uses light from the sun to convert water and carbon dioxide into sugars. That we can minimize plant biology to this one task is a vast oversimplification on our part, while also being a testament to the way that biology balances the chemistry that it performs. When plants are basking in the sun and soaking up carbon dioxide, they are also drinking through their roots and gobbling nutrients from the earth. Photosynthesis is tricky work. The growing plant needs to also have proteins that control this activity, as well as an architecture to support it, which itself requires even more proteins. All of this says nothing of the plant's needs to grow, sense and respond to its environment, and reproduce. (Hope Jahren's excellent memoir, "Lab Girl" details the lives and happenings of plants in entertaining and engaging prose, should you want to know more.[2])

And while the only part we use for brewing is the grain, even this fragment is endlessly complex. The grain is barley's seed and, as such, needs to be able to turn into its own plant someday. The stuff that we care about in the beer making process, starches, and the proteins needed to break them down, come packed with cells that are filled with the information (DNA) and other junk necessary for a growing plant.

There are two kinds of barley that get used in brewing, two-row and six-row, named so for the number of rows of grains growing on each stalk. Two-row barley contains more starch and less protein per grain (including the starch digesting protein, called amylase) than six-row barley. This little bit of information may not seem important now, but is one of those unintended consequences that are useful to understand for the beer making process.

When I say that the grain contains amylase proteins, I am not being entirely truthful. Each seed has the potential to contain amylases, but what they actually must have are the directions for how to make them. Proteins are tricky creatures. They are really useful for performing all sorts of tasks in nature, but they wear down quickly.

Consider a lawn mower. If you could program your lawn mower to constantly drive around your yard, your grass could be perfectly cut. The person who wanted a perfectly manicured lawn at all times might think about doing this. Unfortunately, the lawnmower would constantly be using gasoline and would eventually fall apart. The same is true for proteins, which require lots of energy and break down from overuse. Nature has figured out ways to conserve proteins only for when she actually needs to use them. And, because you can't always just store a protein away somewhere and turn its motor off, nature often doesn't make proteins until it's their time to go to work.

In our grain of barley, the amylase proteins aren't needed until it is time for the seed to grow into a plant, until that growing plant needs the food (sugar) stored away inside of the starch. Nature has also developed ways to take environmental cues to know when to start making these different proteins. In the case of barley seeds, which start to turn into plants (germinate) in the spring, those environmental signals include temperature (18 °C or 65 °F) and humidity. The increasingly warmer and increasingly wetter soil of the spring months, let barley know that it is no longer winter and that it is safe to grow. When the seed feels these triggers, it starts producing lots of proteins, some to break down the starch, some to break down the different parts of the seed, and others to perform all of the other jobs that need doing, including recycling old proteins that are no longer useful.

When we are getting barley ready for the brewing process, we actually make the seed think that it is time to grow into a plant. We take the grains, dried out through the heat of late summer and harvested, and get them just hot enough and just damp enough that they start to sprout. Evolution has ingrained these little grains with the directions for growing and producing new grains of its own. Barley has been following these directions for millennia. They can't help but sprout when we hijack nature's instructions.

These steps are the bare minimum, should you want to use barley to brew beer. You could stop at this point and make a clean, bright and pale lager or ale. But, there's more that you can do to the barley. Increase or decrease the alcohol content. Add some extra color. Change the flavor. Give it a thicker mouthfeel. And this is where a little human-run chemistry comes into play.

Malting is the generic name for the process that prepares grains for brewing. Go to any brewing supply company, and you will find over a hundred different kinds of malted grains for purchase. And, for each kind of malt there is a unique malting process. While the sheer numbers of malted grains can be intimidating (I'm intimidated by them), the different kinds of malting can be grouped into a few distinct classes. When we think about malts in this way, it can make figuring out how to use them a little clearer.

Base malt includes the barley that goes through the germination process that I just described. Beer cannot be made without some amount of base malt thrown into the mix. And, as I mentioned before, beer can be brewed solely from the base malt. These grains have both the starches and the starch cutting proteins needed for beer making. After the germination step, the amylase enzymes will keep chopping the starches into sugars, continuing to push the grain from seed to seedling. So, we pause the biology. The grains are gently kiln-dried, which temporarily deactivates the amylase until the barley gets rehydrated at a later step during brewing.

Crystal malts are a modified grain whose sole purpose is to add fermentable sugar, sugar that yeast can eat and turn into alcohol. After the germination process, barley gets heated to a temperature at which the amylase proteins start working in high gear. Remember, the job of amylase is to turn starch into fermentable sugars, sugars that yeast can chew up and spit out as ethanol. The grains are heated to between 66 and 77 °C (150–170 °F) and amylase goes into overdrive. They cut up as much of the starch as they can get to, although they can't get to all of it. The malts get lightly roasted at temperatures that irreversibly deactivate the amylase proteins by causing them to unfold. These malts add very little to the beer in terms of color or flavor as any sugars that they provide will all be fermented into alcohol.

Roasted malts are just like crystal malts except for, as their name suggests, they get roasted longer. During this step the sugars take part in and are consumed by the Maillard reaction, converted into new flavors and new aromas. The longer the roasting time is, the darker the color becomes. The darker the color becomes, the less fermentable sugar there is. The primary reason for using these malts is to add color and flavor.

Other grains can be malted as well. Oats, rye, and wheat can be treated just like barley and used for brewing. These grains don't

BARLEY & MALTING

Barley is the grain of choice for brewing because it contains a lot of starch and the requisite enzymes that can convert that starch to fermentable sugars

4-row barley
More starch
Less protein

6-row barley
More protein
Less starch

Germination
Seed produces enzymes that prepare it to grow into a plant. Some of these enzymes convert starch to sugar

Base Malt
After a short ferment and a short, drying roast, base malt contains enzymes for mashing; some of the starch has been converted to fermentable sugar

Coloring Malts
The germination and roasting is controlled to give the right amount of proteins and sugars for flavor and color development through the Maillard reaction

Crystal Malt
After a long ferment and a short roast, crystal malt contains fermentable sugars

Germination

Roasting

have nearly the same amylase content as barley and thus won't brew as well on their own. They mostly get used to add interesting flavors to the finished beer. And, finally, adjuncts are grains that get added for the sole purpose of providing starch.

A MONSTER MASH

The grains have been grown and malted. The hops have been harvested and are ready. Even the yeast has been cultured. Farmers and artisans and scientists have done their work, built up their philosophies so you can choose the ones that will best fit your beer.

Your first act as a parent is to decide which of these ingredients to use and how to use them. Daunting, right? I remember with our first child, Erika and I were on edge about everything. Are we feeding her enough? Is she crying? Do we need to go and get her? Is she getting enough sleep? Are we getting enough sleep? (The answer to the last question was a definite, "No.") I'm not really sure how we got by. I'm not really sure how any parents and caregivers do it. I know that we relied on our moms for a good long while. At the start, they were there to give us support and to give us a break when we needed a little rest.

While there is no one quite as wonderful as a grandma while you are starting to brew (seriously, though, nobody is better than a grandma), there are more than enough resources than you can possibly ever use.

The same is true of brewing. I am astounded at the number of blogs and forums and books and clubs that are dedicated solely to brewing. There are plenty of brewing books that are the equivalent of "What to Expect When You're Expecting."[3]

I think looking at ale styles, and different recipes for them, is the most illustrative way to start approaching building a beer recipe.[4] Starting with the lightest, a pale ale. Pale ales have a nice golden color and a malt backbone that isn't overpowering but will provide a compliment to the hops that get used later. For a pale ale you mostly use base malt and add in a bit of lightly roasted, caramel malt. Moving a shade darker to amber ales, keep the same combination as for pale ale but add a bit of darker caramel malt to go along with it. The malt flavoring gets richer and the color moves from golden to red. A brown ale has the same recipe as a pale ale with the exception that you swap out the dark caramel malt for a chocolate malt. A porter uses the amber ale recipe and adds chocolate malt. And, a stout uses the porter recipe with the addition of some roasted barley.

There is an inclination for many to think that porters and stouts have more alcohol than the lighter ales. The dark colors and bold flavor would seem to indicate that the alcohol is also forward. But, this is typically not the case. The intense colors and flavors in the darker malt are only present because the transformation that the sugars go through during roasting (the Maillard reaction). Less sugar means lower alcohol. The alcohol content of any beer can be increased by the addition of some crystal malt, which will only have a small effect on the flavor.

Choosing a malt profile, in my opinion, is the most important step in brewing. These flavors comprise the lasting character of your beer. The other flavors that we associate with beer, the bitter and citrus and floral of the hops and the fruity from the yeast fermentation, are fleeting. As I've got older and tried more beers, I am finding that I immediately notice the malt profile and how it supports and complements the other flavors that get built in later.

When I started brewing, all of the recipes that I used included both syrupy liquid and powdery dried malt extracts. There are a couple of reasons for this. The first is that they are easier to use and a little more foolproof than recipes that only use grains. Really, they are close approximations for using the full-grain recipes. In a large scale, more industrial environment, the sugars and starches and flavors of barley are extracted from the malted gains into hot water. The water is boiled down into syrup or freeze-dried into a powder. When you are ready to brew beer, you just add the syrup and powder to some water and, *voilà*, you're ready to go. The other reason to use extracts for beer making is that all-grain recipes require extra equipment.

I really liked the beer that I made using extracts. They were full and flavorful and always turned out well. But after making a few, I've decided that I want a little more control. Extract brewing leaves no room in any of the malt for subtlety. The way the syrups and powders get made, means they end up with lots of sticky sugar and very little of the more understated flavors that come from careful roasting.

Transitioning from extract to all-grain brewing is like moving from taking care of a baby, to taking care of a toddler. When you're caring for an infant, there are a lot of restless nights because you want to make sure that your baby is safe. Every whimper, every sniffle, every cough sends you running. But, there is lots of snuggling and bonding, as well. When your child becomes a toddler, your worries completely change. First off, you don't worry as much that your child will just randomly stop breathing. (Yes, that sounds bleak. But, as a first time parent I always wondered how anyone would allow ME to take care of a child. I was certain that I was going to do something catastrophically wrong.) You get used to the things that you used to fret over. Toddlerhood brings a new set of responsibilities and a new set of things to watch and worry. But you become much more effective at influencing your child, who themselves have become semi-rational, or at least semi-semi-semi-rational. A toddler starts to develop a real personality, too. They have meaningful physical, mental, and emotional interactions. And, you become more and more comfortable introducing them to new activities.

After getting used to brewing with extracts, you understand some of the ins and outs and realize that even if you totally screw up, your beer is probably going to be more than passable. All-grain brewing gives you the chance to build depth and interesting notes into your malt profile. I still use kits from brewing supply companies, in which all of the grains are pre-selected and which come with very detailed directions. But, as I continue to brew more, I'm sure I'll get to the point where I am comfortable making my own recipes.

The process of extracting the flavors and sugars and polysaccharides from the malted grains and into water is called mashing. If your only responsibility was getting the water to remove these chemicals from the grains, then using the highest possible water temperature would be ideal. Along with flavor extraction, mashing allows the brewer a more nuanced control over the amount of fermentable and non-fermentable sugars. Fermentable sugars are polysaccharides with 3 or less glucose molecules strung together. These are the sugars that will eventually be turned into alcohol, as their name suggests. The liquid that gets produced during mashing tastes sweet because of these sugars. Non-fermentable sugars are polysaccharides with 4 or more glucose molecules strung together. These polysaccharides give your beer a fuller mouthfeel; beer with large polysaccharides is a bit thicker than beer without, which tastes thin.

The balance between large and small polysaccharides is an important one. And, thankfully for us, nature has given us a way to control this balance.

Most of the sugars in base malt (and any adjuncts that get added to the mix) are tied together in long and complex starch molecules. During germination, the amylase enzymes start cutting up starch into sugars that the growing plant can use for food. It turns out that there are two kinds of amylases: α-amylase and β-amylase. α-amylase indiscriminately chops up the starch, giving both large polysaccharides and small, fermentable polysaccharides. β-amylase only cuts off small, fermentable polysaccharides from the end of a starch polymer. α-amylase activity will result in both alcohol and mouthfeel in the finished beer. β-amylase activity will only result in increased alcohol in brewing.

To illustrate how these two enzymes work, I'll give you an example from my own life. My family loves corn on the cob. It's not summer until the first good sweet corn of the year starts showing up. When I eat corn, I'm a slob. I eat from the middle. I eat from the end. I basically just pick the corn up and eat from whatever part is closest to my mouth. Erika is nice and neat. She eats her corn in neat little rows from one end to another. There is an old Disney cartoon with Mickey and Donald and Goofy called "Mickey's Trailer."[5] In it, Mickey and Donald eat their corn to the sound of a typewriter typing. As they get to the end, there is a loud "ding" for the carriage return as the corn gets shifted up so that they can eat the next row of kernels. That's kind of what Erika is like, without the sound effects. Anyway, in this analogy the corn is our starch. Erika is the β-amylase, and I am the α-amylase.

Chemical reactions tend to happen at faster rates as you increase the temperature. The amylase chewing reaction speeds up as you turn up the heat of the stove. There is a point, though, at which protein-run reactions don't go any faster, and actually start slowing down. That is because, as the temperature goes up, the protein unravels. And, when a protein unravels (the fate of all proteins at high temperatures), it loses the ability to do the jobs it is supposed to do. So, when the temperature gets too high, the amylase proteins lose their shape and stop chewing up starch.

Mashing

During mashing, flavors and colors are extracted into the liquid while α-amylase and β-amylase chop up the starch into sugars. α-amylase deactivates at temperatures above 65 °C.

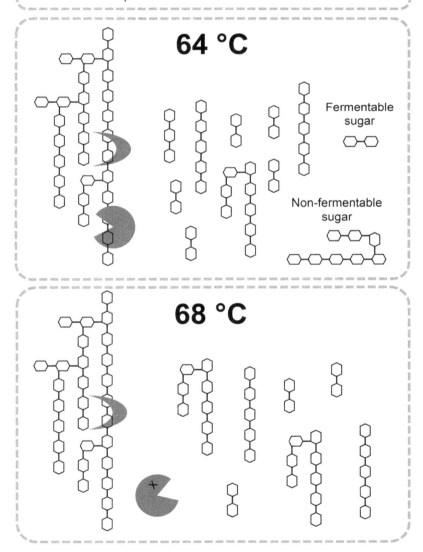

Because every protein is unique, there can be differences in the temperatures that they unfold. α-amylase, it turns out, keeps its structure and activity at higher temperatures than β-amylase. α-amylase starts to unravel at over 70 °C. β-amylase starts to unravel when the temperature gets above 65 °C.

When you go to start brewing, you add our malted grains to water and set the temperature such that you favor the kind of beer that you want. If you want a high alcohol content and are less concerned with mouthfeel, then you bring the liquid to 64 °C. If you want a beer that tastes thicker and are less concerned with the amount of alcohol, then you will bring the liquid to 67 °C. Of course there are all sorts of degrees of changes and outcomes that you can try to get. The point is that if you need to increase the fermentable sugar content, then the temperature has to be just below 65 °C. If, instead, you need to increase the un-fermentable polysaccharide content, the temperature should be just below 70 °C.

Roger Barth is a chemistry professor at West Chester University in Pennsylvania, about a 3-hour drive away from where I live. He teaches a popular class on the chemistry of brewing and has written a really useful book on the subject.[6] Of course I own this book because I love both beer and chemistry. Barth's book was the first place where I learned how to calculate color and alcohol content for any combination of grains. This know-how opens up a world of possibilities for brewing.

After you pick your grain combinations and set a temperature cycle, mashing is a lot like steeping tea. The hot liquid hydrates the grains and allows the amylase enzymes to get at the starches. And, it pulls from the barley all of those flavor and aroma molecules, products of the Maillard reaction. All of this mashing produces a liquid, sweet and rich in color, called the wort.

WHY SO BITTER?

The final step in the human-controlled chemistry of beer production involves the addition of any other flavors, unachievable through the Maillard reaction, that the brewer wants to add. In most cases, this involves the addition of hops.

Hops are the flower of the *Humulus lupulus* plant. Beer makers who brew with hops are going after two components of the flower.

The first is a compound called humulone, which will eventually add a characteristic bitterness to the beer. Brewers in England determined that humulone was a natural antibiotic, which could keep beer from spoiling over long time periods. India pale ales were one of the first beers to employ such large amounts of hopping so that the beer would keep as it travelled from England to the soldiers of the East India Company. The other compounds are lighter, what you might expect from a flower, floral and often citrusy. The difficulty with hops is that getting the bitter and getting the floral, require completely different approaches.

Going back to the parenting analogy. Bribing children only gets you so far. I've gotten to the point in parenthood where I understand the value of a bribe. I also know that they are never truly fruitful; they always come back to haunt you in the end. Give a kid a bribe and you will placate them or get them to focus on something else for a moment. After a while, though, the bribe will just make them unpleasant to deal with. But, normally, I don't care. Bribes are usually meant to give my wife and me a little bit of space to breathe. And, when the kids get nasty again, as they will certainly do, I'll just have to find a new way to bribe them.

Hopping is similar to bribing your kids. After you add the hops to your wort, they immediately start releasing all of those floral and citrusy flavors. (Your kids are sweet after you bribe them.) But, if you want bitter notes to your beer, those hops are going to have to boil in the wort for a long time (around an hour). When the liquid goes bitter again (when the kids start being bratty), that's the time to add some more hops (give another bribe), so at least the bitter has a hint of flower to go with it.

That really is pretty much how adding hops works. (Although there are certainly other ways of doing so.) You add some hops to the wort and bring it up to a boil. The liquid initially extracts humulone, along with many of the essential oils that bring floral aromas with them. As the wort continues to boil, the floral notes evaporate from the liquid, never to return. Over the course of an hour, the humulone goes through a rather dramatic chemical conversion in which nothing is lost from the molecule but some of its parts just change place. The new compound, called isohumulone, is what brings the characteristic bitter flavor to hopped beer.

Hop Chemistry

The citrus and floral aromas from hops are quickly extracted but also evaporate quickly

The bitter flavor from hops only develop after 1 hour of boiling

SENDING THEM OFF

The last step of the brewing process is fermentation. At this point, your beer has pretty much everything that you yourself can give it. There is not much more chemical-human intervention to be had. You turn the wort over to the yeast and hope that they can finish the job that you started. We've come to understand yeasts enough to know, pretty precisely, what they will do in the brewing process. Humans have done some interesting things over the years to select, and now change the yeast, to produce different finished beer properties. Even if we know what we should expect, the process is out of our hands.

And this is the stage that is comparable to sending your kids off to school. You have done your best to prepare them. You kind of expect what they will learn and how they will grow. But, you have to trust in fate that you have given them the foundation that they need.

Having complete control over your beer, and your kids, is kind of boring. Industrial and craft beer producers need to have really tight control so that their product never changes. This is opposed to wine making where the wine changes with the vintage. Commercial beer makers don't have that luxury. But, when you are making beer at home, what you make doesn't come out programmed like some robot. The time and the vagaries of chemistry and biology give your beer a personality that you just can't get from a store. While sometimes those personalities are bad, most often time they are mixed in with too much other goodness.

So, cheers to my mom and dad who knew when to step in and when to stand back. I hope that I can do the same with my own kids. Now, hurry on over here so we can share a beer.

REFERENCES

1. C. Papazian, *The Complete Joy of Homebrewing*, Harper Resource, New York City, , 3rd edn, 2003.
2. H. Jahren, *Lab Girl*, Alfred A. Knopf, New York City, 2016.
3. H. Murkoff and S. Mazel, *What to Expect When You're Expecting*, Workman, New York City, 2008.
4. J. Palmer, *How to Brew: Developing your own Recipes*, [document on the Internet], HowToBrew.com; [cited 2016 July 5], available from: http://howtobrew.com/book/section-4/experiment/developing-your-own-recipes.
5. *Mickey's Trailer*, Hollywood, Disney, 1938, Film, director Ben Sharpsteen.
6. R. Barth, *The Chemistry of Beer: The Science in the Suds*, Wiley, Hoboken, New Jersey, 2013.

CHAPTER 14

Cocktails

When the season starts to turn from winter to spring, one of the things I most look forward to is enjoying a gin and tonic.

   I don't think there is another drink that is as crisp and refreshing, the perfect counterpoint to a sunny day. While there are many cocktails that are much fussier to make, the gin and tonic packs surprises that outpass its unpretentious production. Key to all of this, for me, is that the flavor is something very different to either gin or tonic on its own. As many of my friends already know, I am no big fan of gin or tonic.[1] I find tonic water overly bitter for my preferences, even when sweetened. Gin just doesn't do it for me either. It has always tasted heavy and penetrating to me. Even smelling it, the oils and aromas coat the inside of my nose and sinuses. The sensation is similar in my mouth, lingering long after the liquid has passed into my stomach. The persistence of it all feels weighty to me.

   But together, a gin and tonic is something very different to the sum of its parts. The flavor is bright and crisp, the bitterness of the tonic and the juniper-dominated flavors of the gin have become the most interesting part of the drink. "Composing" gets thrown around in the culinary world for the way different dishes are put together. For plated dishes, to me composition is reflective of writing music. The different instruments, and the notes that they bring, are layered and arranged, creating a sonic experience that builds up about a new whole. The thing about cooking is that I like most of the ingredients by themselves, much like I enjoy listening to instrumental solos. I liken mixing cocktails more to the composition involved in painting. Individually, canvas and paints hold no real aesthetic draw to me. But, a beautiful painting is nothing more than the expert application of paints to canvas, drawing contrast between the colors through shading and direction in a way that evokes emotion. For gin and tonic, these two hues take on more meaning when they mixed together.

Gin is nothing more than flavored vodka, which itself contains only ethanol and water. Vodka starts out with a similar foundation to beer. Food-based starches are broken down and converted into glucose and other monosaccharides. The starches that eventually turn into vodka can come from a number of sources: corn, barley, rye, or potatoes. The monosaccharides then become food for yeasts, which spit out carbon dioxide and ethanol, the byproducts of their digestion. Yes, we drink yeast poop. The difference between beer and liquors, like vodka and whiskey, is distillation.

Beer checks in at an average of around 5% alcohol by volume. Vodka does so at 40%, and you can buy grain alcohol that hits an ethanol content of roughly 98%. To concentrate that yeast waste, it must be distilled.

By definition, distill can mean to isolate the meaning or essence of. For the alchemists who perfected the art of distillation, they used it as a means of extracting the truest parts of a substance, leaving behind its baser and less desirable elements. When they distilled mixtures of dilute alcohol in water, the liquid that they produced had surprising powers.[2] It could ease pain. It was thought to cure disease. And, it could preserve animal tissue. That meats and flesh would not spoil while soaking in this liquid, inspired the practitioners to call it *aqua vitae*, or water of life; things soaked in ethanol were preserved from a normal death. The French later translated *aqua vita* to *eau de vie*, which is still used as a name for many brandies. At any rate, the alchemists thought that they had found the noblest essence of fermented grains, isolated through distillation.

We have since come to realize what happens to liquid mixtures, on a molecular level, as they are distilled. Distillation operates on the foundation that pure liquids have unique boiling points. Water boils at 100 °C. Ethanol boils at 78 °C. Evaporation itself occurs when individual molecules gain enough energy to break free of the interactions that hold them together in a liquid. Heat water enough and eventually a single molecule will gain enough energy, and move around with enough speed, that it is able to fly off into the air, the other water molecules are unable to grasp it strongly enough to prevent it's departure. The fact that water boils at a higher temperature than ethanol, indicates that the interactions between two water molecules are stronger than the interactions between two molecules of ethanol.

In a mixture of ethanol and water, the ethanol will boil before the water does. If you keep the temperature low enough, well below 100 °C, you can extract quite a bit of the ethanol from that mixture, by trapping the ethanol vapors as they leave the liquid. The problem with water and ethanol is that they don't neatly distill. There is a strong interaction between these two molecules. So, you can never really get absolutely pure ethanol from a mixture with water through distillation. And, for the vodka (and other hard liquors) that we drink, we never distill from a liquid that is a simple combination of water and ethanol. There is all sorts of chemical gunk left over from the grains and yeasts that the ethanol and water are distilled from. Some of this gunk will eventually be useful, like it is for aged rums or whiskeys. But, for vodka, the water and ethanol need to be separated from this mess. (Potato vodkas are historically known for the oils that are difficult to get rid of through distillation.) For many brands of vodka this usually means multiple distillation steps until the unwanted stuff has been left behind; sometimes it takes a while to find the right "essence".

A clean vodka is the starting point for any gin. Gin is vodka that is flavored. Juniper berries constitute the major flavorant, but other botanicals can be used as well. These ingredients can include: angelica root, bay leaf, cardamom, citrus peel, coriander, cubeb, fennel, ginger, grains of paradise, lavender, and orris root. Each of these brings their own subtle flavors to gin. (One of the failings of the English language is that our words cannot capture the nuance between different flavors. Our language broadly describes many flavors at once.) Juniper berries lend several aromas (that come from specific chemicals): pine (pinene), camphor and pine (camphene), woody and peppery (sabinene), rosemary (cineole), sweet citrus (terpinene), woody and spicy (cymene), and menthol and woody (terpinen-4-ol). These aromas are complemented by the other ingredients in the gin.

These molecules are what make gin different from vodka. They are the reason for gin's flavor. They are the reason why I experience gin coating the inside of my mouth. And, much of the way these molecules act (from their aromas to how they need to be extracted, to how they affect the flavor of gin and tonic) can be gathered from their molecular structures.

While all gin is flavored with come combination of botanicals, the specific types of gins are distinguished by the methods in which they are favored. Gin labeled as "gin", with no other descriptors, is vodka that has been flavored with botanicals and through the precise addition of very specific aroma chemicals. For "gin", just steeping these ingredients in the vodka is enough to achieve the desired flavors. Distilled gin is vodka that is redistilled with juniper and several botanicals. After distillation, other flavorings can be added. Distilling the vodka with the botanicals completely changes the profile of the molecules that are extracted. The higher temperatures in the distillation enable the removal of more molecules than can be removed by steeping the ingredients at room temperature. London dry gin is vodka that is redistilled with juniper and botanicals with no other additives. The flavor profiles are slightly different in all of these. And, each brand of gin will have slightly different flavors in the gin and tonic.

A "tonic" is literally a liquid that is meant to tone and heal the body. The "tonic" that we refer to in gin and tonic, is a carbonated drink that contains quinine. I have always found the history of quinine to be really interesting. Naturally, quinine is found in the bark of the cinchona tree, which is native to Peru.[3] Peruvians had been using this bark to treat a disease whose symptoms are similar to those of malaria. An Italian Jesuit observed this in the late 1500's, and Europeans have been using it as a treatment for malaria ever since. Eventually, they figured out how to extract quinine from the bark and isolate it in powdered form. British soldiers with the East India Company, which also gave us India pale ales, were ordered to take quinine to prevent catching malaria, and mixed the powder in water with syrup to cut the bitterness. This medicinal tonic is the direct ancestor to the tonic water we have today, in which a quinine and sugar syrup is added to carbonated water.

Gin and Tonic

A gin and tonic is a seemingly simple drink. It is clear and clean, not muddled by unnecessary flavors or textures. 50 mL (1 3/4 ounces) of gin, 93 mL (3 1/4 ounces) of tonic water, and a squeeze of lime are enough to bring me happiness. Playing with variations on this theme, some bartenders and mixologists are bringing the gin and tonic to new places. Adam Bernbach, bar director at the restaurant Estadio in Washington, DC, has crafted a number of unique tonics for their gin and tonic menu. These include an orange-thyme, an elderflower-citrus, a rhubarb-cardamom, and basil tonics. When I visited Estadio, I passed on the dessert so that I could have another G&T.

Dave Arnold, owner and mixologist of Booker and Dax (in New York City) and kitchen scientist extraordinaire, is the author of "Liquid Intelligence", required reading for anyone who is interested in cocktails or science or the science of cocktails.[4] I would wager that Arnold has thought about gin and tonics more deeply than anyone else on the planet. He spends a full 15 pages of his book describing how he has experimented with and how he currently makes his signature drink. He obsesses over bubbles and how to preserve carbonation during drink making. He describes how he makes his own quinine syrup, combines it with gin and water and carbonates the whole lot. As he dispenses this concoction into the serving glass, he also adds a bit of clarified lime juice. All of his perfection is aimed at defining and finding balance.

Whether you get to try one of these high-end versions, or are enjoying one of your own creation at home, a good gin and tonic has a couple of chemically-driven processes going on that enhance the overall experience.

Arnold is right to keep his focus on carbonation. The carbonation is pleasing in the way that it tickles your tongue. The carbon dioxide makes the water a little more acidic, which changes how you taste the drink. But the most important part is the effervescence. Bringing the drink to your lips, the bubbles pepper your face with miniscule drops of water. The way it plays on your skin, before you even smell or taste it, affects your enjoyment of the drink. The carbonation cools your skin and sets the tone for your experience. Once it's in your mouth, each pop of a bubble sends forth a gin and tonic perfume. The surface of a bubble is a hydrophobic environment. Some of the aromatics in gin and tonics will naturally find their way to the border between bubble and water. As the bubble lets loose, these aromatics fly off to your aroma receptors, where you experience these flavors at the whim of the carbonation. The bubbles are like a time-release capsule, delivering just enough botanicals to optimize your enjoyment, keeping plenty in reserve so that your mouth and sinuses are not overwhelmed.

The most striking thing to me though, is how much the flavor of gin and tonic differs from that of just gin and just tonic. A gin and tonic isn't a binary combination to our palate where 1 plus 2 equals 3. It is an integrated experience in which 1 plus 2 equals spoon. The gin and tonic change each other. Well, they collectively change your perception of each of them. This experience isn't unique to gin and tonic. But an inspection of gin and tonic flavor can be illustrative for how we experience other foods.

Words (at least my words) won't do justice to an experience like this. Thankfully, Pixar has gone to the work of turning these feelings into art for me. There is a scene in *Ratatouille* where the main character, Remy, takes a bite out of a cheese.[5] The flavors manifest themselves as filled circles and comma-shapes, an

BUBBLES AND FLAVOR

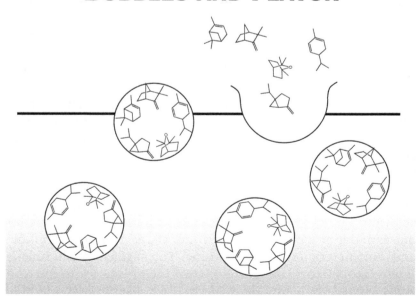

earthy yellow color, gently flitting in and out of the screen. Next, a bite of strawberry elicits open circles, sharp purple lines, growing and popping, waves floating by. Taking a taste of both the cheese and strawberry together, Remy experiences bursts of red, new patterns emerging and fading in and out of his consciousness. It really is a beautiful scene. And it captures in images, a translation of our emotional response to food; a task I could never hope to accomplish with my writing.

I also know that chemistry alone can never account for these experiences. But there are chemical foundations upon which these flavor experiences and emotions are built.

Flavor itself is a complex combination of taste and smell, with taste occurring on the tongue and comprising primarily of sweet, salty, sour, bitter, and savory (umami). These "simple" sensations aren't directly linked to our brains. That means that sweet doesn't just hit our nerves, and that's the end of it. The presence of sugar can depress the way we experience bitter and salty and sour. Salt can depress bitter and enhance savory. (Arnold suggests adding drops of saline to most cocktails as a taste and flavor enhancer.) Tonic water contains sugar for the sole reason of minimizing the bitter flavor of quinine. When gin gets added to the mix, the sugar certainly minimizes its bitterness as well.

But, what I'm most interested in is the way we experience the flavor of the botanicals and how it is affected by the quinine in the tonic water.

The experience of aroma is translated from odor molecules to your brain through one specific type of protein: G protein-coupled receptors. Research on these proteins was awarded the Nobel Prize in 2012.[6] As nature loves systems that work well, these types of proteins are found all over our bodies. They are used to: activate our immune system, regulate our mood, control our blood pressure, and define several of our senses (sight, taste, and smell), among other purposes. The basic premise of how all of these proteins work goes as follows. The protein bridges a cell membrane; one end pokes itself outside of the cell, and the other end remains firmly within it. The part that lives outside has a single job, it waits around for the right kind of molecule to come along and nestle into the protein. When this happens, the part of the protein that is inside of the cell changes its shape, setting off a cascade of signals and eliciting some biological response. These proteins are notoriously tricky to study. My colleague, Stefano Costanzi, has done some incredible work predicting what the outside portions of different G protein-coupled receptors look like and, further predicting what kinds of molecules will come along and bind to them.

For a long time, chemists and biologists would describe the way smaller molecules would fit inside of larger proteins using a lock and key metaphor. Only one key design will fit the lock on your front door and open it. In this telling, the protein was designed by nature to perfectly fit a specific small molecule. Only this one molecule could fit. And, in fitting, only this one molecule could provoke a biological response.

But, this isn't completely accurate. These systems are more like a hand and glove. The glove (protein) is flexible and can flex to accommodate a number of different hand sizes (small molecules). Even if that hand is missing a finger, it will still fit into the glove.

If we look at our gin and tonic aroma molecules and think about how they operate in this scheme. Camphene has an aroma similar to that of camphor, which is a lot of what you are smelling when you are near a eucalyptus tree. Camphene has a sharp, bright, and woody smell to it, not unlike a citronella candle. There is a G protein-coupled receptor that provides a home to camphene when it enters your nasal cavity. Once camphene finds its way into the protein, its insides change shape and send a signal to your brain that you are smelling something that is sharp, bright and woody.

There are several chemical processes that modulate the intensity with which we smell camphor that are directly related to how we experience gin and tonic. Camphene doesn't necessarily find its protein glove right away. In fact, the protein won't allow camphene in, unless there is a critical mass of camphene molecules hanging around in the nasal cavity. This has to do with how snugly the hand fits within the glove. If the glove finds an ideal match, it won't need to have a lot of suitors waiting around. A three fingered hand will still fit the glove, but there needs to be a lot of three fingered hands floating around before the glove accepts that fate. This is reflective of the thermodynamics of small molecule binding to a protein.

Flavor Biochemistry

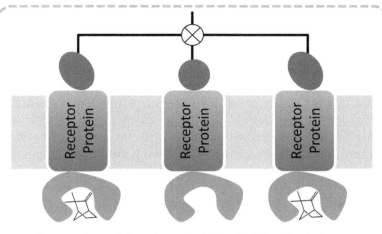

Our experience of flavor is predicated by the interaction between flavor molecules and receptor proteins

Some of the factors that affect the interaction between a flavor molecule and a receptor protein

The strength of the flavor-protein interaction

The strength of the flavor-saliva interaction

How other flavor molecules interact with the protein

How many flavor molecules are nearby

Interactions with other proteins

The strength of the flavor molecules' interactions

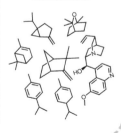

In addition to this, the pace at which camphene slides into the protein can also dictate the biological response (*i.e.* the way we experience aroma). My kids have gloves that they wear to play in the snow during winter. Those gloves have straps around them to keep their hands dry. When they get dressed to go outside, my kids are impatient. They think they can force their hands into their gloves even though the straps have tightened the entry of the gloves. They push and push and push and eventually get their hands in. Just like my kids and their gloves, camphene and its protein receptor also have to contort themselves a bit before camphene can find its way in. And, just as there are kinetics for entry, there are kinetics for leaving a protein as well.

These simplifications (yes, they are simplifications) are muddled by the way we experience aroma: from a complex system of proteins and molecules. But, the complexity is what gives rise to the pay-off that we experience when we taste a gin and tonic.

There are roughly 400 known kinds of odor receptor proteins and thousands of known odor molecules. Each protein has the potential to bind lots of different molecules (likely not at one time, though). Concurrently, each individual molecule can find its way into several different protein receptors. The protein (glove) that binds to camphene (hand), can also fit other molecules in the gin and tonic mixture. The individual protein might elicit a single biological response (aroma). But the intensity of that response is tempered by the strength and the speed at which a particular molecule fits into it. Camphene binding will elicit an aroma from one protein that lies somewhere on the intensity continuum of that protein. Camphene will do the same thing with other odor receptors in the nasal cavity as well.

When you drink gin, there is an integrated biological response that you experience. The flavor of gin is the muddled happenings of all of the ins and outs of all of the different gin molecules going into all of the different proteins that they fit into. When you add tonic (and specifically quinine) to that mix, you invariably change the way that the gin botanicals find their ways to the odor receptors. For gin alone, the strength at which a protein grabs onto camphene takes into account the stability of the camphene in the liquid (gin plus saliva) and the stability of the camphene in the protein. The difference of these two stabilities is what drives the camphene to interact with the protein instead of just hanging out in your saliva. When you add quinine to the mix, the camphene will want to interact with that too. The ability of camphene to evoke an odorous response is the difference between the stability of camphene in the protein (this hasn't changed) and the stability of camphene in the liquid in your mouth (this has changed and now must include the way that camphene interacts with quinine).

Way back in the vinaigrette chapter, I talked about how molecules that look alike tend to be attracted to one another. There is no small resemblance between camphene and one side of quinine. That doesn't mean that the two will be inseparable. What it does mean is that they will interact enough to change the way that camphene interacts with the odor-receptor proteins. And, this isn't just true for camphene. It applies to the other gin botanicals as well. Adding quinine to gin shifts the way that the botanicals interact with the proteins in our nasal passageways and, because of that, also changes the overall flavor of those botanicals in gin and tonic.

The simplicity of cocktails lays bare their reliance on the way mixtures of aroma molecules shift and alter how we actually experience their individual flavors. While foods can provide substance and sharp contrasts in texture that affect how we enjoy eating a particular dish, cocktails mostly need the interplay between the aroma molecules to create encounters that are appealing.

And so we go to great distances to find and extract aroma molecules into our liquors. Amy Stewart is an author who has written both fiction and nonfiction books. The focus of her nonfiction is squarely aimed at the natural world where she has written about bugs but has mostly focused on plants. She has a delightful book titled, "The Drunken Botanist." In this book she chronicles our human history of using plants to make alcohol. My favorite sections are when she describes the plants (their roots, their stems, their leaves, their flowers, and their seeds) that are used to extract different flavors.

Historically, the use of alcohol to extract flavors was the realm of apothecarians and alchemists. Before the advent of modern medicine, when plants provided the only source for relief from an ailment, tinctures were made that captured the essence of these plants using ethanol. Some of these tinctures can get fairly complicated.

This is certainly the case of Chartreuse. Chartreuse is a liqueur made by Carthusian monks in the mountains outside of Grenoble, France. I was in Grenoble for a research conference in 2013, and one of the planned excursions was a trip to the monastery where the monks lived and to the distillery where the liqueur was blended and aged in underground cellars. The monastery was striking. Remote from any cities or towns, I can understand how a person could find peace and contemplation in that surrounding. Remote from cities and towns, I cannot imagine how the monks who have lived there for several hundred years ever carried the raw materials that were used to build the monastery. The buildings are as beautiful as the surroundings. And these surroundings provided the source of all of the herbs and plants that go into making Chartreuse. Over 100 different plants are used to make this liqueur, many of which are a closely held secret. During the tour we were told that only 2 monks at any time have access to the full recipe, and when they die, the recipe is passed on to younger members of their order. The secrecy and the silence of these monks really plays into the intrigue surrounding Chartreuse. But, their dedication to this craft really is on display in the caves where Chartreuse is aged. Huge wooden barrels store vast quantities of the Chartreuse mixture, each barrel indicating a blend that has been aged for a different amount of time. One of the things that the tour guides did fill us in on was that there is no age when Chartreuse is perfectly ready. Chartreuse only becomes ready when a master blender tastes it and deems its aging to be complete.

On my tour were roughly 60 or so chemists, who, like me, had a million questions that were not allowed to be asked. How did they make the tincture for each individual plant? Did the blending occur all at once or were some tinctures added later? What is the biggest difference between green Chartreuse and yellow Chartreuse?

Sadly, none of our questions were answered. The cone of secrecy lives on. But, we did get to try a number of different Chartreuse blends. In the small group of chemists whom I had known before the tour, we all agreed that the taste of neat Chartreuse didn't really do it for us. It was too medicinal, which shouldn't be a surprise as that's what it was once used for. Even walking through the caverns where Chartreuse was aged, the air was thick with its aroma. The tour was really like wafting through a haze of Chartreuse.

In my opinion, the essence of Chartreuse is not to stand on its own, although it certainly does that. The essence of Chartreuse is to lend its legion of aroma molecules to the complexity of a cocktail. Just like the gin botanicals find a pleasant complexity when mixed with quinine, gin and Chartreuse botanicals can also be blended into something enjoyable. The Last Word is one of these cocktails.[3] Combine equal parts of gin, green Chartreuse, maraschino liqueur, and lime juice [22.5 mL (3/4 of an ounce) of each will give one drink], into the bowl of a cocktail shaker with ice. Stir gently (with a spoon or by swirling the shaker with your hand). I like using a metal shaker for making drinks because the metal, which rapidly conducts heat, allows me to quickly estimate the temperature of the liquid inside as it cools down. When the shaker has taken on a layer of condensation and does not feel like it is getting any colder, you are finished.

Dave Arnold, the brilliant obsessive that he is, has published his guidelines for making and mixing cocktails in "Liquid Intelligence."[4] He argues that only drinks that are supposed to have foams should be shaken; these tend to be drinks made with milk or egg white. When other non-foaming cocktails are shaken, the agitation adds air bubbles and ice chips to a drink that is supposed to be crystal clear. I tend to agree with him, and not James Bond, on this shaken *versus* stirred debate. Shaking a cocktail will cool it faster, but the end product isn't nearly as appealing to the eye. And, while some claim that shaking "bruises" the botanicals in a cocktail (oxidizes them and ruins their flavor), I'm not sure that there is much evidence in support of that. Though the botanicals may not actually be ruined, the presence of bubbles, on their own, can impact the way you experience the flavor of the cocktail. (I will discuss this in more detail in the chapter on ice cream.)

The blending of floral and citrus botanicals seems to be highlighted in gin-based cocktails more than those of other liquors. The Aviation is a marriage of gin with maraschino liqueur and *crème de violette*.[3] In the States, maraschino automatically means cherries that have been stripped of their natural flavors and filled with red (or green) dye and pumped full of grenadine-like syrup. Real maraschino comes from sour Marasca cherries. The liqueur has a slightly bitter flavor and the aroma of almonds. *Crème de violette* is the sweetened essence of Parma violets (or other "sweet" violets). It is a beautiful blue color, which translates nicely to The Aviation. To make this drink, add 60 mL (2 ounces of gin), 15 mL (1/2 an ounce) of Maraschino liqueur, 7.5 mL (1/4 of an ounce) of *crème de violette*, and 22.5 mL (3/4 of an ounce) of fresh lemon juice to a cocktail shaker filled with ice. Gently stir or swirl until the shaker does not feel as though it is getting any colder. Strain and serve. The proportion of these ingredients can be altered to match your tastes, as can the garnish, which often includes a real maraschino cherry.

How anyone found this combination of ingredients is beyond me. I can't fathom being thrown behind a bar, with all of its different liquors and liqueurs and juices and be asked to come up with something completely new. Where would you even start with a process like that?

Thankfully there are a few guidelines that can get used.

In "Liquid Intelligence," Dave Arnold spends an entire chapter discussing his database of cocktails, and, specifically, his analysis for trends that could be found from all of the recipes.[4] What he came up with was a broad generalization for different target dilution (how much ice should melt), alcohol content, sugar content, and acid content. The human palate is a tricky thing. What we like or dislike cannot necessarily be predicted by chemical or biological tools. There is more of a direct line from what we find pleasing to our cultural influences, than there is to any chemistry or biology or neuroscience. So, what Arnold has done is rather ingenious. Analyzing the things that we like (cocktails that have become ingrained in any bartenders repertoire), he has found the balance of the different chemical components that are the essence of these recipes. That is, the question he has asked is, "Has culture defined what a balanced drink is?" I'll go over his analysis of stirred drinks, which are the majority of the drink recipes that I am discussing in this chapter. He found that most drinks called for between 90 and 97 mL (3–3 1/2 ounces) of liquid in the recipe, with an initial alcohol content between 29 and 43% by volume. Important to stirred drinks are some sugar and some acid. In the Aviation, the acid comes from the citrus juice and the sugar comes from the *crème de violette*. The stirring with ice alters all of these levels. The dilution of water brings the alcohol and sugar and acid levels to a range that we find pleasing. In his book Arnold uses these guidelines in a number of interesting ways. He shows how he follows the guidelines when designing a new cocktail. He also employs them to make bottled versions of his favorite cocktails, taking into account the dilution, sugar, and acidity of the final drink.

But even with these simple recommendations how do you know which of the 1000 bottles will actually pair well? This is where trial and error comes in. When you have done enough cocktail making, you develop an understanding of flavor profiles. So, you look for familiar notes in a liqueur that may be new to you and you substitute it into a recipe that you already know. I do this a lot when cooking. If I buy an ingredient that I've never used before, the first thing I do is taste it. After I have a feeling for its flavor, I can use it to make something familiar, something that I know I like.

Much of haute cuisine and haute mixology, though, is pushing the boundaries of familiar. There is an interesting theory floating around in the culinary world that has a chemical focus to it. It goes like this: foods that share the same molecules will taste good together. Martin Lersch, PhD chemist and purveyor of the blog *Khymos*, which looks to understand and push the modernist cuisine movement, has a whole set of posts organized around this theory. His series "They Go Really Well Together" was meant to explore strange food parings based on shared molecules.[7] Now, not just any molecule could be used. Cellulose is in every plant-based food. Lersch was looking for molecules that played a major role in the flavor profile of the food. In an example of this, Lersch and his readers explored recipes that combined strawberries and coriander, which both contain (Z)-3-hexanal, a molecule with a "green" or "vegetal" aroma. Lersch came up with a strawberry and coriander foam for serving with ice cream. Other reader-submitted recipes ranged from a strawberry and cilantro terrine to a strawberry and cilantro guacamole, where the strawberries substituted for tomatoes.

One of the reasons why this theory has caught some traction is that not only are the combinations unusual, but the results, like

FLAVOR PAIRINGS

a good cocktail, are more than just a combination of ingredients. When ingredients share major odorant molecules, the dish that they make doesn't quite taste like the binary addition of those ingredients. The flavors are different. Going a little deeper into how we experience flavors, we can start to understand how these flavor pairings create such surprising dishes.

When we eat a strawberry, we don't experience each individual flavor molecule in the strawberry (the primary ones being 3-hexenal, furanone, methyl butanoate, ethyl butanoate, methyl 2-methylpropanoate, or 2,3-butanedione). We don't take a bite and say, "Oooh, I can really taste the fruity ethyl butanoate in this," or, "This strawberry has some really buttery 2,3-butanedione going on." No, most of us eat a strawberry and taste "strawberry". Flavor is an integrated experience. While we don't get each individual flavor, all of those flavors are necessary for us to recognize the food as "strawberry". We have a learned and ingrained the definition in our unconscious minds for what chemicals (and their concentrations) are necessary to elicit the "strawberry" flavor. Remove any one of those, and strawberry will taste just a bit different.

This is not to say that we are incapable of tasting and recognizing individual chemicals; it is possible to retrain our brain to do so. Master Scotch whisky blenders are able to derive the same flavor profile of their blended scotches year after year using the same single malt scotches every time. However, the single malts vary in their flavor profile from one year to the next. The master blenders have to figure out how to adjust their ratios at each step in the blending process to get just the right mix. To do this, they must be able to sense the absence of specific molecules in one single malt and recognize its presence in another.

When mixing strawberry and coriander, the (Z)-3-hexanal is an important part of the integrated experience for each of these foods individually. When you combine the two, however, the context that you experience (Z)-3-hexanal has now changed. Because of this, the flavor profile of the strawberry shifts away from "strawberry" and the flavor profile of "coriander" shifts away from coriander. Our brain gets a little confused, is unable to fully recognize either ingredient, though they are familiar, and is surprised (and hopefully delighted) by the combination.

A collaborative team including the programmers at IBM and the Institute of Culinary Education (including chef's James Briscione and Michael Laskonis), used IBM's supercomputer, Watson, to find unusual food pairings.[8] In a similar way to what Dave Arnold did with his cocktail database, Watson had thousands of recipes inputted into its memory. The IBM programmers had Watson search for recipes that featured a single ingredient. Once these were collected, Watson found ingredients from these recipes that were not found as combinations in any other recipe. The point of the search was to find foods that aren't often paired together but that might actually still taste really good. In the process of doing this, Watson would spit out odd ingredient combinations and the ICE culinary team would try to devise recipes that utilized these. They even tried their hands at a couple cocktails and came up with recipes that used meat: "Hoof-n-Honey Ale" and "Shrimp Cocktail" as well as some more traditional cocktails: "Ivorian Bourbon Punch" and a "Japanese Wasabi Cocktail."

Because not all of us have the benefit of chemical composition data for all of our ingredients, or the help of a supercomputer, there is always just plain old trial and error sprinkled with a dash of creative intuition. The most surprising drink that I have ever had was a Penicillin. At first, I really couldn't believe that I was ordering it, but it sounded so bizarre that I just had to try it. A Penicillin, created by bartender Sam Ross, is a scotch-based cocktail.[9] Now, I do love scotch. But when I think of cocktails, usually those cocktails are made with whiskey and not whisky. To make a Penicillin, start by muddling some fresh ginger in the bottom of a cocktail shaker. Add 60 mL (2 ounces) of blended scotch, 22.5 mL (3/4 of an ounce) of lemon juice, 22.5 mL (3/4 of an ounce) of honey syrup (equal parts honey and water), and ice. Shake vigorously to thoroughly, mix the ginger and chill the drink. Strain through a fine mesh strainer. Top the drink with a thin layer of Islay scotch. Ginger and honey and scotch. Who would have thunk it?

Even if you're not up for being surprised by your cocktails, the traditional standbys never fail to please.

If spring ushers in gin and tonic, fall brings Manhattans. This is another simple drink. 2 ounces of rye and an ounce of sweet vermouth are chilled by stirring with ice in a cocktail maker. Strain the ice and add a few dashes of bitters. The woody barkiness of the vermouth and bitters make the drink smooth and warm. The Manhattan was the drink that taught me the importance of measurement in drink making. Previously, I had just sort of eyeballed things. My Manhattans became good when I actually added each liquid in the correct proportion.

Another crowd-pleaser in the summer are my mojitos. I was introduced to mojitos by Mike, the bartender at the faculty club at the California Institute of Technology. Lime, sugar, mint and rum perfectly blended for a perfect drink. When my wife and I moved from California to Maryland and our front yard flower patch sprouted more mint than we could handle, I knew we had to figure out how to make mojitos. After playing around with a few different recipes, and I finally cobbled a few together that would make a pitcher's worth. This recipe uses 2 1/2 limes (cut into 20 wedges) and makes 4 drinks because that's how many we usually need at one time when we have guests over. Muddle 5 of the lime wedges with 110 grams (7 heaping tablespoons) of sugar and the leaves from 5 sprigs of mint. Add the remaining 15 wedges and muddle until all of the lime juice has been released. Add 225 mL (7 1/2 ounces) of white rum and mix to suspend the sugar. Add 600 mL (2 1/2 cups) of club soda. Pour over ice in 4 highball glasses and enjoy.

I wrote earlier that cocktails are at the mercy of flavor pairings more than other types of food. And, while this is true, the perfect mixtures can seamlessly capture the essence of time and space. The essence of botanicals reflecting the essence of any emotion. There is chemistry there. But, I'd say that the cocktails are the purpose and product of the alchemy from ages past.

REFERENCES

1. M. Hartings, The Best Science Writing Online 2012, in *I Love Gin and Tonics*, ed. B. Zivkovic and J. Ouellette, Scientific American/Farrar, Straus, and Giroux, New York City, 2012.
2. A. Rogers, *Proof*, Houghton Mifflin Harcourt, New York City, 2014.
3. A. Stewart, *The Drunken Botanist*, North Carolina: Algonquin Books of Chapel Hill, Chapel Hill, 2013.
4. D. Arnold, *Liquid Intelligence: The Art and Science of the Perfect Cocktail*, W. W. Norton, New York City, 2014.
5. *Ratatouille*, Hollywood, Pixar, 2007, Film, directors Brad Bird and Jan Pnkava.
6. The Royal Swedish Academy of Sciences, *Press Release: The Nobel Prize in Chemistry 2012* [document on the Internet]. NobelPrize.org; 2012 October 10 [cited 2016 July 5], available from: https://www.nobelprize.org/nobel_prizes/chemistry/laureates/2012/press.html.
7. M. Lersch, *They Go Really Well Together* [document on the Internet]. Khymos; [cited 2016 July 5], available from: http://blog.khymos.org/tgrwt/.
8. IBM and Institute of Culinary Education, *Cognitive Cooking with Chef Watson: Recipes for Innovation from IBM & the Institute of Culinary Education*, Naperville, Illinois: Sourcebooks, 2015.
9. P. Clarke, *Penicillin Cocktail Recipe* [document on the Internet]. Serious Eats; [cited 2016 July 5], available from: http://www.seriouseats.com/recipes/2009/11/time-for-a-drink-penicillin-cocktail.html.

CHAPTER 15

Ice Cream

The first thing I ever really got good at making is ice cream. The Cuisinart ice cream maker that my wife and I got as a wedding gift might have been the present that I was most excited about. The place settings and napkin rings were nice and all, but the ice cream maker held unlimited possibilities! And we have just about tried them all. I have made orange sorbet, pumpkin ice cream, candied ginger ice cream, and everything in between.

The first ice cream that I made was a vanilla bean custard ice cream that came in a recipe book that we got with the ice cream maker. We used vanilla bean because we had never actually used a real vanilla bean before. We wanted to feel adventurous even though we were just making vanilla ice cream. And, we wanted something that we were pretty sure we couldn't screw up. What came out on the other side of our efforts blew us away. This was quality ice cream that could be *quenelled*, more of a solid custard than any ice cream than I had ever eaten.

In the time since, I have come to embrace the utility of a good base vanilla ice cream. A recipe like this gives you a foundation to build upon, should you want to change to something else. The first base recipe I used wasn't even a vanilla ice cream recipe. It was Ina Garten's espresso ice cream recipe from *Barefoot Contessa Family Style*.[1] I liked this recipe so much that I figured I could retrofit it to almost anything I wanted, which I could. The base that I took from this is as follows. Heat 720 mL (3 cups) of half and half milk until it steams. While the cream is warming up, beat 6 egg yolks, 67 grams (2/3 of a cup) of sugar, and a pinch of salt, with a mixer until the eggs turn a light yellow color. Slowly add the hot cream to this while stirring with the mixer. Return the liquid to the pan and heat until the liquid coats the back of a wooden spoon. Strain the custard through a sieve and refrigerate until completely chilled. Freeze the custard in an ice cream maker.

This recipe held me in very good stead for a while. I was able to tweak it when and how I wanted. And it was simple enough to use.

About three years ago, though, I discovered David Lebovitz. Well, I certainly didn't discover him in the literal sense. The man worked at Chez Pannise and Zuni Café before he started writing. I mean that I was finally exposed to his work and his recipes. There are so many wonderful voices – thoughtful, introspective, and artistic – in the food and cooking world that it is hard to work your way through all of them. This recipe is enough to make me grateful to have found David's work. But, his most recent book *My Paris Kitchen* is a beautiful fusion of French and Moroccan cooking, complementing his wonderful writing style.[2] I really do recommend it, much like this ice cream recipe, which I use for ice cream that I make.[3] Heat 250 milliliters (1 cup) of whole milk, a pinch of salt, and 150 grams (3/4 of a cup) of sugar in a saucepan. Scrape the seeds from a vanilla bean with a paring knife and add them to the milk along with the rest of the bean. Remove the pan from the heat, cover, and let the mixture steep for an hour. Prepare an ice bath to pre-cool the heavy cream. I put copious ice and water into a large ceramic bowl. I add 500 mL (2 cups) of heavy cream to a 4-cup measuring cup, and then place the cup gently into the ice bath, using the cups handle over the edge of the bowl to prop it up. In another bowl, mix the egg yolks together until homogeneous. Reheat the milk and add some of it to the eggs with stirring. Add the egg–milk mixture back to the saucepan. Cook over low heat until the custard coats the back of the spoon. There are two indications that I use to know when the custard is about ready, and when it is ready. The first indication is that any bubbles in the pan start to die down. Whipping the eggs yolks together naturally forms a stable foam. That foam persists even when you add the milk. As you start heating the custard, the bubbles of the foam will increase in size and then start disappearing. This is your cue that the custard is just about ready. After that, really wait until the custard coats the spoon. When you draw a finger down the wooden spoon, the custard should be plainly visible, holding its shape. Experience really helps you to have a good feel for what this looks like. When the custard is ready, pass it through a sieve into the chilled cup that contains the heavy cream. Chill the custard and freeze in an ice cream maker.

The other author that I trust on all matters of ice cream is Max Fakowitz, who used to do lots of ice cream writing for the blog *Serious Eats*.[4] He has some really great posts on the topic, including one of my favorites titled, "5 Ice Cream Myths that Need to Disappear."

When I wrote earlier that ice cream was one of the first dishes that I got good at making, part of that is to say that the ice cream that I made was really good. But, then again, what ice cream isn't. What I meant by this statement is that I gained a pretty decent feel for what is going on in each step of the process and knew how changes in the recipe necessitated other changes along the way.

In short, I've developed my own philosophy for ice cream making.

To best describe this, I think it would be best if I start at the end: freezing.

FREEZING

The first homemade ice cream that I can remember having as a kid came out of one of those wooden bucket ice cream makers that you dumped rock salt and ice down the side of. The cream and sugar would go into a little tub that fit nicely packed in among the salt and ice. These things had hand cranks on them to churn the cream as it froze.

Looking at pictures of these now, I'm taken back to summer nights playing with my cousins. We would be catching fireflies or shooting off fireworks or shooting bottle rockets at each other. We'd take a break to have a bowl of homemade ice cream. I always thought this stuff was more reminiscent of frozen soup than ice cream. But that didn't matter. It was a perfect treat for those hot and muggy summer nights. I liked mine plain, but I recall most of my cousins dumping Hershey's syrup on theirs and stirring it all up (which really turned it into a soup, a chocolatey soup, but a soup nonetheless). As a kid, I don't think I ever realized who was running the crank to churn this ice cream for us. I was pretty oblivious to everything. But, I suppose now I can send a thank you out to whichever of my parents or aunts or uncles did all of that work while we got to run around being kids.

The texture of this hand-cranked ice cream isn't what I'm really aiming for when I'm making it at home. The crunchy soupy nature of the stuff just doesn't cut it for me. But, that description "crunchy soupy" does get at the most important aspect that needs to be controlled throughout every step in the ice cream-making process. "Crunchy" implies that there are noticeable chunks of ice forming. "Crunchy" is not what I want in an ice cream (even though I'll gladly eat it like I did when I was a kid). No, ice cream should be smooth. Ice cream should be creamy, obviously. And, one of the main keys to smooth and creamy ice cream is rapid freezing.

Ice freezing is a kinetic process as much as it is a thermodynamic process. That is, we most easily think of the thermodynamics of ice freezing. We know that we have to cool down liquid water to turn it into ice (thermodynamics). We know that water freezes at 0 °C (thermodynamics). We know that freezing water transfers energy to its surroundings (thermodynamics). But, what we don't often think of is that while water is freezing, the liquid water is still moving (kinetics). And, if you freeze water slowly, you'll make larger ice crystals (kinetics). But, if you freeze water quickly, you'll make smaller ice crystals (kinetics). Faster freezing equals smaller ice crystals, equals smoother ice cream.

So, how do I make the smoothest ice cream ever?

That's easy.

Just use liquid nitrogen, which checks in at a temperature just below −196 °C (−321 °F).

Liquid nitrogen gets collected from the air by one of two ways. Air is condensed around chillers (think of the way water condenses around a cold glass on a hot and humid day) and then distilled to isolate pure nitrogen. This distillation works the same way that alcohol distillation works, just at much lower temperatures. In distillation, a mixture of liquids is heated. Because each component of the mixture has a different boiling point, you can purify and isolate each part by slowly boiling off each chemical, one at a time. Liquid nitrogen is also collected from air by compressing it with huge pistons and allowing all of that gas to rapidly expand. The physics of gas expansion causes rapid cooling. If the expansion is properly controlled, to just the right conditions, liquid nitrogen is produced.

Either way, liquid nitrogen is one of those laboratory chemicals that really delights. Though it can be dangerous (the intense cold is obviously a hazard as is the potential for asphyxiation when too much liquid nitrogen evaporates in a closed area), there are ways to reduce the risk by properly handling it. Only trained professionals, who understand how to reduce risk and know how to respond if anything does go wrong, should be using it or overseeing its use. (Even if you don't work in a lab, like me, and can't get your hands on any liquid nitrogen, you can still find places that serve liquid nitrogen ice cream. Here in the greater D.C. area we have Nicecream Factory, located in Arlington, Virginia. If you're in the UK, Chin Chin Labs operates their liquid nitrogen ice cream shop out of London.).

I get to make liquid nitrogen ice cream for the students who take my chemistry of cooking class. Sadly, I can't make it at home. The sight of billowing smoke wafting over a bowl as you constantly add liquid nitrogen is something to behold. (Yes, I am aware that it's not really smoke; it's condensed water from the air. But smoke is a much more evocative and lyrical word).

I let my students get in on the act too. All through the semester, I ask for volunteers to help me with different cooking demonstrations during class. And, all year long it's like pulling teeth to convince people to come up to the front of class and help me. But, when I ask if people want to help pour liquid nitrogen, there is not enough room for everyone who wants to help. For the lucky few who do get to come up, I give them a few quick safety tips (making them wear proper protective gear, showing them how to hold the liquid nitrogen Dewar, giving them pointers on how fast to pour the liquid nitrogen) and off we go. I put the ice cream base into a bowl and start churning the cream with an electric mixer. The student starts pouring in liquid nitrogen. And, 10 minutes later, we have perfectly frozen, perfectly creamy ice cream.

The second law of thermodynamics tells us that heat will always flow from a hot item to a cold item. And so it is with our liquid nitrogen ice cream. Because of the extreme difference in temperatures (−196 °C for the nitrogen and somewhere between 4 °C and 20 °C for the cream) there is lots of heat that can flow from the cream to the nitrogen. That heat flow away from the cream to the nitrogen chills and eventually freezes the cream.

César Vega, whom I discussed in the egg chapter, got his PhD studying ice cream science at University College Cork, in Ireland. Along with a group of video producers, he has put together several of kitchen science videos that nicely illustrate what happens to all of the chemicals involved in making ice cream.[5] In these videos, César also gives a few very important tips for making good ice cream. It is vital, he says, to cool the cream to between −5 °C and −7 °C. So that should be your target when you put your cream into your ice cream maker, or my goal when I start dumping liquid nitrogen onto my cream.

Max Falkowitz had a really nice post on his *Serious Eats* blog that shows all of the ways that ice cream made in professional kitchens is better than what we can prepare at home.[6] One of the biggest differences, he says, is having a freezer that is dedicated for ice cream storage that can reach very cold temperatures. Vega agrees with this. In his tips, he says that one of the most important things that you can do is put your ice cream in a hardening room (a special freezer that circulates really cold air to quickly finish cooling your ice cream). After you have churned your ice cream, it's still not quite ready yet. It has to cool down a little farther. Vega says that somewhere between only 30 and 40% of the water actually freezes into ice crystals. When your ice cream goes into the freezer, you're not quite at that 30–40%. So most of the water molecules are still moving around. This finishing step also needs to happen fast; fast freezing means the water molecules don't have a chance to move very far before they freeze, resulting in smaller ice crystals.

While most of us don't have access to freezers that are dedicated and modified for the sole purpose of making ice cream, we do have access to different heat exchangers at home. The importance of these for heat flow is illustrated in the difference between liquid nitrogen ice cream, and ice cream made in the wooden bucket with the rock salt and ice heat exchanger. The freakishly low temperatures of liquid nitrogen demand fast heat flow from the cream. The rock salt-ice set up can't get anywhere near these rates.

It seems strange, but it takes energy to melt ice. This seems counterintuitive to us because ice will spontaneously melt when left on a countertop. The spontaneous nature of this process (something that we observe just happening) belies the fact that the ice is pulling in energy from the environment while it melts. Just as ice melting is a key step in cooling down cocktails (the different concoctions pushing its own energy into the ice cubes in a cocktail shaker), the flow of heat from cream to the rock salt-ice mixture is what freezes the cream. Rock salt plays an important role too.

Most of us instinctively know that salt water freezes at a lower temperature than pure water. The salt trucks that go out in force to prepare our streets for an impending winter storm are an admission of that. The rock salt is there to keep water from freezing on top of the pavement. In chemistry, we talk about this in terms of colligative properties. Translated, what that means is that the more stuff you put into water (salt, antifreeze, sugar, *etc.*) the lower the freezing point becomes. As these changes are driven by the absolute number of things you put into water (not the mass of stuff you put into water), some additives have more effect than others. Sodium chloride has extra power for lowering the freezing temperature because it will dissociate into a sodium ion and a chloride ion. Compared to propylene glycol, which does not dissociate into multiple ions when it is in water, salt has a greater colligative effect on water's freezing point. Also, salt can be a generic term, describing any ionic compound, that encompasses more than just table salt. Calcium chloride (containing one calcium ion and two chloride ions) and magnesium chloride (containing one magnesium ion and two chloride ions) are two other "salts". When these chemicals dissolve in water, they dissociate into three things and, logically, have a larger effect on the freezing temperature of water than sodium chloride.

At any rate, when the ice melts in our heat exchanger it reaches a lower temperature, because of the salt, than ice on its own would reach. This low temperature provides another heat transfer step for the cream; after the heat from the cream melts the ice, more heat is required to increase the temperature of the resulting salt water.

The heat exchangers in home ice cream makers, like the one that I have, are filled with either propylene glycol (the stuff in antifreeze) or some sort of brine (mixture of water and a salt, not necessarily just sodium chloride). The benefit of these heat exchangers, over the rock salt-ice heat exchangers, is that the brine or propylene glycol–water mixture are pre-cooled together. The rock salt-ice heat exchanger would be closer to my Cuisinart if you had first made salt water and then frozen that into ice cubes.

It is somewhat ironic that you have to melt ice (in a heat exchanger) to make ice (in your ice cream). But, such is the balance that is demanded by chemistry. The second law of thermodynamics states that energy can neither be created nor destroyed. The energy that melts ice has to come from somewhere. In our ice cream, it just so happens that the energy comes from water freezing. We play chemical games so that the flow of energy naturally goes to the rock salt-ice or to the antifreeze or to the liquid nitrogen. But there is poetry in the transformation of water to ice, acting at a distance, causing a foreign patch of ice to melt into a pool of water.

And, truly, the freezing of ice is itself a beautiful process.

That all snowflakes are unique is not necessarily a fact, but they are no less beautiful because of that. I have fond memories of making paper snowflakes when I was a kid. I now have fond memories of my own children cutting paper snowflakes.

Kenneth Libbrecht, a physicist at the California Institute of Technology, study's the formation of ice crystals as part of his research. Seeing the art in this research he wrote a book, "The Snowflake: Winter's Secret Beauty" that prominently features incredible pictures of single snowflakes, taken by the photographer, Patricia Rasmussen.[7] The pictures themselves are mesmerizing. But the easy beauty of the images is coupled to Libbrecht's careful and caring description of the work that he finds joy in. Artistic expression is often the best way to teach and to learn. While I am used to reading scientific papers, Libbrecht's book has influenced the way I think about my own research projects where I study the way that proteins enable individual gold atoms to join up into small particles, which are then wrapped up by the same proteins that start the pile.

When water molecules freeze, they stop moving. No individual molecule freezes on its own. It needs partner molecules to join in with the freezing process. So, several molecules, while reducing their motions, end up in a slow dance that is joined by more and more water molecules. As part of this waltz, the water molecules take up very specific positions on the dance floor. Remember how the oxygen in water is more negatively charged than the two hydrogen atoms, which have their own slight positive charge. As the molecules join the dance, the oxygen from one water will find the hydrogen from a second water, and their positive and negative charges will grab hold of one another. Eventually, when the dance floor starts to fill up, the partners are all lined up in a shape that is reminiscent of a hexagon. Scientists can actually predict this shape, understanding how the oxygen atoms and hydrogen atoms will search each other out to find the most stable (lowest energy) arrangement.

This hexagonal arrangement holds its form and translates out as the ice crystal forms. For this reason, snowflakes themselves are hexagonal. Well, they're not exactly perfect hexagons. In chemistry and math and physics, we say that they have 6-fold symmetry. They may not have 6 sides. But they certainly have 6 main arms, which themselves have arms that branch out, each branch a mirror image of a branch on an entirely different arm.

So, we can look at a snowflake, as beautiful as it is, and realize how the individual water molecules in that snowflake have arranged themselves.

Beautiful.

Snowflakes don't have a monopoly on this property. Minerals also have larger structures that reflect atomic arrangements. The hexagonal columns at the Giant's Causeway in Northern Ireland are a direct product of the way the individual metal and oxygen ions packed together over 50 million years ago. Sodium chloride forms cubes because its ions line up in a cubic structure. Crystals of gypsum are prismatic because of the type of monoclinic arrangement of the calcium and sulfate ions, along with a few water molecules. These atoms hold their positions so strongly that their geometries can be faithfully replicated billions and billions and billions of times over. Their shapes mean so much. Take a trip to any natural history museum; you'll never look at minerals the same way.

Ice Cream

CHURNING

In our backwards trek through the steps of making ice cream, we come to the churning. Churning the ice cream serves several purposes. It mixes the custard to evenly cool the whole liquid, rather than just the edges. Churning scrapes any ice off of the side of the heat exchanger, keeping the ice crystals as small as possible. Finally, churning incorporates air into the ice cream.

We really don't think about it, but ice cream is a foam, just like shaving cream, spray insulation, and those pads you put on top of your mattress. A foam is made when pockets of air get trapped inside of a solid or a liquid. I think of foams as having an industrial connotation. Styrofoam and dish sponges immediately come to mind.

But foams are important to the culinary world too. Of these, popcorn is one of my favorites. Popcorn starts out as a bunch of starch (amylose and hemi-amylose) and a little bit of water, tightly packed inside of a dense shell. As each kernel of corn is a seed, the starch is there to act as food for a growing seedling. There are several things about popcorn that set it apart from other types

FREEZING WATER

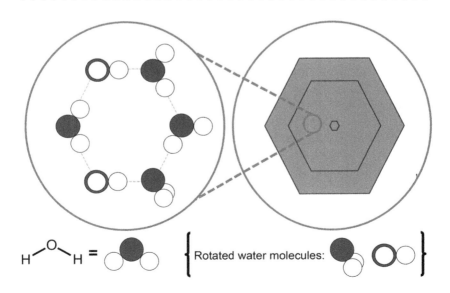

of corn (notably sweet corn, which we eat, and field corn, which is what livestock eat and that can be turned into corn flour, corn syrup, and whisky). Popcorn has more starch and more water than other kinds of corn. And, it also has a denser shell. The dense shell is absolutely critical for making popcorn. As you heat the kernels of corn up, the water reaches such a high temperature that it hydrates the starch and gelatinizes it. Gelatinization is a misplaced name (there is no gelatin in popcorn or other starches) that just means that the individual polymers of starch don't hang on to each other anymore. This is physically manifested in the way that the starch changes from rough and hard, to soft and easy to chew. Eventually the water inside of the kernel reaches such a high temperature that the shell can't contain it anymore. The shell ruptures, and the gelatinized starches expel out rapidly, propelled by the steam. As starch pushes out, it rapidly cools. In the end, you are left with a foam in which a solid network of starch traps air bubbles.

The presence of air within the popcorn is equally as important as the starch is to your enjoyment while eating it. You can do this experiment yourself. The next time you eat popcorn, slowly compress one piece until it flattens out, displacing all the air. Your smashed piece of popcorn will be tough and more difficult to chew, not nearly as pleasing as a properly puffed kernel.

There are liquid foams that we're used to as well. The milk foam on top of a cappuccino is one example. Whipped egg whites with sugar are another example of a liquid foam, which, after it cooks and becomes a merengue, turns into a solid foam.

In each case the presence of air bubbles within the solid or liquid mesh, completely change the culinary properties. The most obvious of these is that foams are softer and easier to eat than the same foods where the air has been removed. Beside popcorn, another good example of this is bread. Bread with a proper crumb (well spaced and distributed air pockets) is delightfully chewy, while dense bread with no evidence of leavening is much more difficult to eat. Foams have taken center stage in a lot of modernist cuisine offerings as well. Many of the food chemicals that get used (xanthan gum and lecithin, for instance) help to stabilize foams made with flavored liquids. The search for new, vegan-friendly food chemicals has even mined the liquid that chickpeas get canned in. Referred to as *aquafaba*, many people have been excited because this liquid has many culinary (and chemical and physical) properties that are similar to egg whites.[8]

Ice cream is no different from other foams. It benefits from the presence of bubbles, which make it airier and effectively taste creamier. Edy's (which also goes by the brand name Dreyer's) is a popular commercial ice cream brand in the United States. They market one of their varieties as "Slow Churned" and boast that it contains half the fat content of their normal ice cream. This is a brilliant and useful and slightly devious marketing scheme. The fact that the ice cream uses half the fat as normal ice cream is a good thing for people on restricted diets. They get away with not adding fat, which would normally add to the softness of the ice cream, by pumping extra air into it. What is devious about this is that cream can be expensive for ice cream manufacturers to purchase. And, by using less cream, it doesn't cost as much to produce a tub of ice cream.

A foam doesn't just form by pumping air into a material (solid or liquid). There need to be molecules that help support the presence of the air bubbles. For popcorn, the starch polymers frame the network of air pockets. In bread, gluten acts as a balloon, trapping carbon dioxide in the rising dough. For egg whites, proteins are also responsible for this work.

For ice cream, the particles of milkfat end up being responsible for stabilizing the foam, and it does so in a sort of round about way. If you let raw milk sit, the cream will eventually separate to the top of the container, the milkfat drawn upward by its lower density and overall hydrophobicity. To prevent this, milk is homogenized, a process that forces the casein particles to glom onto the particles of milkfat. The casein particles maintain the emulsion of milkfat dispersed in water. When milk gets used to make ice cream, there is usually some sort of emulsifier that is added (lecithin and proteins from egg yolks, in the case of custard based ice creams, or hydrocolloids, like carrageenan, for non-custard based ice creams). The emulsifiers disrupt the interactions between casein and the particles of milkfat.

As the ice cream base is cooled, the milkfat, freed of its casein stabilizers, undergoes its own sort of crystallization. It's not a real crystallization, in the strictest sense of the word; a crystal is a well-ordered arrangement of molecules. The fat turns into a glass, a disordered arrangement of fat molecules. As part of this glass transition, little spikes of fat grow out of the particle so that they look more like the end of a mace than the smooth blob that they once were.

When the cream base is churned, air is incorporated into the cream and the fat particles start creating the network that holds these air bubbles in place. As the milk fat gets pushed around during the churning, eventually two fat particles will meet up. The stabby ends of their mace-like shapes will pierce each other's bodies, and the two particles will stick to each other. This happens over and over again until a wide network of speared fat particles forms. The grid of fat particles is set by the size and spacing of the air bubbles that have been incorporated into the cream.

All of the air infusion inescapably leads to an expansion of the ice cream. This expansion is called the overrun. The overrun is no joke. I remember making strawberry ice cream once. All of the cream fit pretty neatly into the ice cream maker. But, I didn't account for the overrun. The ice cream overflowed from the container and got into the gears, jamming them, and ruining the ice cream maker.

FOAM STABILIZATION

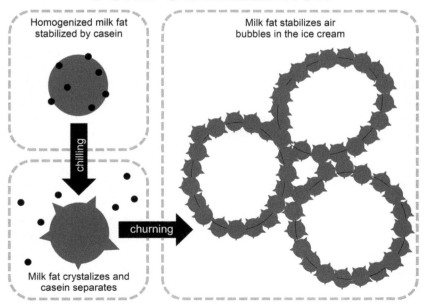

Ice Cream

THE ICE CREAM BASE

And now we arrive at the beginning, making the ice cream base. I suppose our trip through ice cream was kind of like *The Curious Case of Benjamin Button*.[9] When we start at the end, sometimes we can better see what's most important, and our beginnings can accurately reflect our inner character. Organic chemists and pharmaceutical chemists often perform similar tasks called retrosyntheses. In a retrosynthesis, you start with the molecule you are trying to make and, going backwards, figure out the best way to make it. Again, there is clarity for what needs to be done when you take this approach.

The clarity that we find when we start with freezing is that we need to do all we can to keep ice crystals small. We need to prevent the formation of large ice crystals.

Thankfully, there are several (delicious) ways to do this when making ice cream. It helps to think of the ice cream as an emulsion, which it is. If you can further stabilize the emulsion, water molecules will be prevented from separating from the fat particles. If the water molecules are slow to separate, they will form small ice crystals.

Just as in the freezing step, the ice cream base can help to dictate the kinetics of ice crystal formation.

The way that I approach this most often is by making a custard. The egg yolks (at least 5 large egg yolks) supplement the emulsifiers in the ice cream by adding proteins (around 2.7 grams per yolk) and lecithin (around 0.126 grams per yolk),[10] the amphiphilic emulsifier responsible for the stability of mayonnaise. The added sugar plays two roles. First, it slows down the water molecules (just as it does when making jams). Second, large amounts of sugar also lowers the freezing point of the water in the ice cream. In a good ice cream, only 30–40% of the water molecules are trapped in ice crystals. The sugar supports the ice cream in reaching that target.

The ability of the custard to slow down water molecules is the primary reason that it is crucial to heat it to just the right consistency. As I noted earlier in the chapter, I heat the custard until it thoroughly coats the back of the spoon. Part of the thickness is due to water evaporation. Most of the thickness is due the formation of a network of unfolded egg proteins (better described in the egg chapter) that serve as fences to slow water motion.

I find custards to be my favorite option when making ice cream at home. The custard gives me the results I want, with respect to ice crystal size.

The game is a little bit different when I am making liquid nitrogen ice cream at my university. In this case, the freezing is so fast that I don't need the same amount of emulsifiers that I otherwise would. So, for my liquid nitrogen ice cream I make a Philadelphia-style recipe. Philadelphia-style ice cream, also referred to as standard ice cream, doesn't use any eggs in its recipe. The base that I use is as follows. Thoroughly combine 480 mL (2 cups) of half and half milk, 340 grams (1–12 ounce can) of evaporated milk, 150 grams (3/4 of a cup) of sugar, and 5 mL (1 teaspoon) of vanilla extract until the sugar dissolves. Some heating may be necessary for this. Cool the liquid before freezing.

This recipe benefits from the addition of evaporated milk, which contains carrageenan. Carageenan is a polysaccharide hydrocolloid that is extracted from several kinds of seaweed. They have strong interactions with proteins, which is why they pair so well with milk.

While many people would be comfortable adding evaporated milk to their ice cream, there are just as many who would balk at the option of adding straight carrageenan to the mix. This is plainly silly. There are lots of natural chemical stabilizers that can be used to enhance your ice cream. Eggs (with their proteins and lecithin) absolutely constitute a chemical stabilizer. Some shy away because "chemicals are bad," and some shy away because mass-produced ice cream uses non-egg stabilizers. But, as Max Falkowitz has so eloquently put it, "Stop bashing ice cream stabilizers!"[6] Depending on the type of ice cream that you want to make, what you want its texture and flavor to be, there are a number of stabilizers (carrageenan, gelatin, xanthan gum, guar gum, and others) that get used by professional craft ice cream makers. So, you shouldn't be limited either! There are plenty of people who share there secrets on-line. Go and search them out, or do your own experiments at home!

MAKING NOT ICE CREAM

Speaking of branching out, I like to branch out in my frozen-treat making, too. For a while, Erika and I would make sorbet much more often than ice cream, with mango sorbet being one of our go-to recipes. The following includes a number of my favorites.

So, what's the difference between ice cream and gelato? You will certainly get a different answer if you ask an Italian. I have had my fair share of gelati over the course of the two weeks I have spent in Italy on separate vacations. (When my daughter was younger, we had a book called "Olivia Goes to Venice" about a young pig who travels with her parents to Venice and eats gelato twice a day before accidentally demolishing the tower at San Marco.[11] My visits weren't much different, other than the tower destruction.) But my answer to the gelato *versus* ice cream question lies in subtleties more than radical departures from one style of recipe to the others.

The biggest difference comes from the temperatures at which they are stored for serving. Gelato needs to be kept at a much higher temperature than ice cream. For both recipes, the amounts of fat and the amount of eggs span a similar range for ice cream and gelato. The other big difference is the amount of overrun. While some ice creams can have up to 100% overrun (doubling in size because of the amount of air incorporated), gelato's overrun shouldn't exceed 10%. The result is a more dense dessert served at a temperature that allows you to experience the intensities of flavors and aromas found within the cream.

My favorite at home gelato recipe is from Marcella Hazan's *Essentials of Italian Cooking*.[12] Her strawberry gelato is a thing of beauty. To me, it sings summer time. And it is beautifully simple to make. Puree 225 grams (1/2 of a pound) of strawberries in a food processor with 150 grams (3/4 of a cup) of sugar and 180 mL (3/4 of a cup) of water. Whip 60 mL (1/4 of a cup) of very cold heavy whipping cream until slightly thickened. Fold the cream into the strawberry puree and freeze in an ice cream maker. When my family and I are going to eat this right away, I cut down on the sugar and the water using only 1/4 of a cup of each (60 mL of water and 50 grams of sugar).

When I want to cut the cream out and just enjoy the flavor of pure, (nearly) unadulterated fruit, I make sorbet. David Lebovitz has a wonderful citrus sorbet recipe that is even easier than the strawberry gelato recipe, but no less wonderful.[13] Juice some citrus. (I like to use oranges, but this works for any citrus fruit). For every 250 mL of juice, use 50 grams (1/4 of a cup) of sugar. Put the sugar in a small sauce pan and add just enough of the juice to cover the sugar. Heat until the sugar dissolves. Add to the reserved juice and freeze in an ice cream maker. I usually make this recipe for my students, especially for those that might be vegan or lactose intolerant. While I make it to cater to a select group of students, it ends up being a showstopper that the whole class enjoys.

The final recipe that I will share is for how I make coconut ice cream. I love coconut ice cream. It may be my favorite of any ice cream flavor. The key to this recipe, which I learned from David Lebovitz, is toasting the shredded coconut.[3] It really heightens the flavor, making it taste more coconut-y. Also, I substitute coconut cream (for the heavy cream) and coconut milk (for the whole milk) of my standard ice cream recipe base that I use. To keep it completely vegan, I also omit the eggs. Cans of coconut milk and coconut cream often come with guar gum or some other stabilizer. Having these stabilizers around is necessary for a smooth ice cream. In a medium pan, toast 240 mL (1 cup) of dried, shredded coconut (unsweetened) until it takes on a nice golden color. Add the coconut (keep ¼ of the toasted coconut in reserve for later), the coconut milk, and 150 grams (3/4 of a cup) of sugar to a small pot and heat until the sugar dissolves. Place a lid on the pot and remove from the heat, allowing the milk to extract the toasted coconut flavors over about an hour. Strain the milk into a 1 liter or 4-cup measuring cup containing the coconut cream. Freeze in the ice cream maker. The reserved shredded coconut can be placed in the ice cream at the end of churning or used as a garnish when you serve the ice cream.

REFERENCES

1. I. Garten, *Barefoot Contessa Family Style: Easy Ideas and Recipes that Make Everyone Feel Like Family*, Clarkson Potter, New York City, 2002.
2. D. Lebovitz, *My Paris Kitchen: Recipes and Stories*, Ten Speed Press, Berkeley, 2014.
3. D. Lebovitz, *The Perfect Scoop: Ice Creams, Sorbets, Granitas, and Sweet Accompaniments*, Ten Speed Press, Berkeley, 2010.
4. M. Falkowitz, *Scooped* [document on the Internet]. Serious Eats; [cited 2016 July 5], available from: http://sweets.seriouseats.com/scooped.

5. C. Vega and AnimationKitchen, *AnimationKitchen's videos* [document on the Internet]. Vimeo; [cited 2016 July 5], available from: https://vimeo.com/user18092612/videos.
6. M. Falkowitz, *Secret Tools and Tricks of the Ice Cream Pros: How to Make Creamy Ice Cream.* [document on the Internet]. Serious Eats; [cited 2016 July 5], available from: http://sweets.seriouseats.com/2014/03/secret-tools-and-tricks-of-the-ice-cream-pros-chefs-secrets.html.
7. K. Libbrecht, *The Snowflake*, Voyageur Press, Minneapolis, 2003.
8. *The Official Aquafaba Website* [document on the Internet]. Aquafaba.com; [cited 2016 July 5], available from: http://aquafaba.com.
9. *The Curious Case of Benjamin Button*, Hollywood, Warner Brothers, 2008, Film, director David Fincher.
10. H. McGee, *On Food and Cooking*, Scribner, New York City, 2004.
11. I. Falconer, *Olivia Goes to Venice*, Atheneum Books for Young Readers, New York City, 2010.
12. M. Hazan, *Essentials of Classic Italian Cooking*, Alfred A. Knopf, New York City, 1992.
13. D. Levovitz, *Blood Orange Sorbet Recipe* [document on the Internet]. DavidLebovitz.com; [cited 2016 July 5], available from: http://www.davidlebovitz.com/2008/02/blood-orange-so/.

CHAPTER 16

Pie

Some recipes are classical.

They get passed down from generation to generation, ensuring that the food they love will continue to be enjoyed. Developing new dishes is hard enough. Making certain that someone else can replicate what you've done makes it even more difficult. So, we tell our children. We tell our students and apprentices. We tell our friends. They go off and put our lessons to use. They figure out what works and what doesn't. They make improvements. They refine. Eventually the recipe doesn't really look the same as it did when you first came up with it. But, what remains is what the alchemists deemed, "the essence." It could be a flavor combination. It could be some bit of science that has optimized how a dish is made. All of these optimizations add up, and your recipe is perfected by the collected wisdom of history.

All of our cookbooks are filled with the knowledge of the ages. Most times we don't realize it. Our ancestors have created some incredible kitchen chemistry. In many ways, this book serves as a testament to that. While I do enjoy speculating how certain foods came about, often I am just in awe at the precision of those cooks who came before us.

Chocolate mousse is a perfect example of this.

I had my first real mousse at the home of my friend Lionel Cheruzel. Lionel is another former colleague from Caltech. He and I were both postdocs in the same lab. It's a strange position to be in. As a postdoc, you are hoping to springboard yourself into a faculty job at another university. But, there are only so many openings and there are so, so many postdocs working in high profile labs all over the country. There aren't enough jobs to go around. Naturally, Lionel and I were both applying for the same jobs at the same time. There was an unavoidable sense of competition. But, we were (and are) friends. It was a stressful time for both of us.

In the time that we've known each other, we have both shared meals and foods that are important to us. He was very gracious to compliment the bread that I served him. (For a Frenchman to say he likes your bread is about as high of an accolade that I could hope to get.) And, Lionel shared his chocolate mousse with us.

I still remember one particular evening that my burgeoning family spent with Lionel. This is when we first met the woman he would marry (and have twins with), a fantastic scientist, in her own right. I remember that our oldest, Kaitlyn, who was just starting to really learn how to talk, kept saying, "*Bon jour*. Good bye." But, what I most recall is Lionel's chocolate mousse.

Being an American, I don't think I'd ever really had chocolate mousse before. Really, what I probably had was some sort of custard. The richness of a real mousse is undeniable. Lionel told me that it was his grandfather's recipe and that he wouldn't dare share it (wisdom of the ages is special, you know). So, to approximate, I'll consult Julia Child, who has been America's official translator of French cuisine since the 1960s.[1] [Beat 4 egg yolks with 150 grams (3/4 of a cup) of sugar until pale yellow. Child writes, "... until it falls back upon itself forming a slowly dissolving ribbon." Beat in some orange liqueur and place the mixing bowl on top of a, near boiling, pot of water. Continue mixing until it foams and becomes hot. Then place the mixing bowl into cold water and beat again until it has the consistency of mayonnaise. In a separate bowl, gently melt 168 grams (6 ounces) of chocolate and then beat into the egg yolk and sugar mixture. Beat 4 egg whites until they form a soft peak. Add 15 grams (1 tablespoon) of sugar and continue beating until they form stiff peaks. Fold some of the egg whites into the chocolate, egg yolk, sugar mixture until incorporated. Fold the rest of the egg whites into this mixture. Portion out into dishes and refrigerate.]

I love the detail of Child's recipe here. Beating the mixture until it forms a ribbon that slowly dissolves into the surface. It's really an exquisite bit of observation of a scientific process. The egg proteins need to unfold and start making a foam. The properties of the foam, specifically its viscosity, need to match that of a gently dissolving ribbon. And, this thickness is controlled by egg protein unfolding and aggregation. The whipped egg whites are added back as a way to incorporate more air into the mousse, making it feel light even though its laden with richness.

I'd like to step back for a moment and admire the chemical beauty of this recipe. The original author, whomever that may have been, needed to have an immense amount of chemical intuition at their fingertips. They needed to know how egg yolks transform with whipping and heat. They needed to know how to temper chocolate. They needed to know that egg yolks have the requisite amphiphilic emulsifiers that could facilitate the incorporation of chocolate. They needed to know that whipped egg whites provide a skeletal structure to many foods. All of these steps are pure chemistry, whether their creators realized it or not. And each little step, every single stage of protein unfolding and aggregation, brings us one of the finest desserts ever crafted. Historical dissemination of chemical wisdom, passed from generation to generation.

Chocolate mousse has always seemed very exotic to me. Perhaps it's not to Lionel. The precision of the directions, perhaps, is probably what keeps this recipe out of the standard repertoire of the average American cook.

But just because a recipe comes off as mysterious does not mean there is no less science involved. To illustrate this, I'd like to talk about chocolate chip cookies. Nestlé Toll House chocolate chip cookies, to be precise.

Chocolate chip cookies are pretty standard fare for any kid growing up in the States. The original recipe was developed at the Toll House Inn in Massachusetts in the 1930's[2] and is published on the back of every package of Nestlé's chocolate morsels. [Combine 315 grams (2 1/4 cups) of all-purpose flour, 8 grams (1 teaspoon) of baking soda, and 8 grams (1 teaspoon) of salt. In a separate bowl beat 230 grams (2 sticks) of softened butter, 150 grams (3/4 of a cup) of sugar, 165 grams (3/4 of a cup) of brown sugar, and 5 mL (1 teaspoon) of vanilla extract until creamy and continue to beat while adding 2 eggs. Gradually mix in the flour mixture and then 340 grams (1 12-ounce package of chocolate chips). Form into balls and bake for 10 minutes at 190 °C (375 °F).]

Not only do most kids grow up eating these cookies, their recipe is also one of the first ones that kids get to help out with in the kitchen. Adding ingredients, stirring everything together, eating left-over dough (even though we probably shouldn't) are all memories that are pretty engrained from my childhood.

You would think that because this recipe is so codified, it would be pretty foolproof. For the most part it is. But it seems like everyone makes these cookies a little differently. Maybe it's how you measure ingredients. Maybe it's your oven. Maybe it's the cookie sheet that you use to bake them on. Nobody's Toll House chocolate chip cookies are ever the same.

And, for my money, nobody makes them better than my sister, Johanna. Erika and I consider ourselves to be better than competent in the kitchen. And, of the two of us, Erika is the baker. But neither of us can make chocolate chip cookies like Johanna. I knew I had to talk to her about her "secrets" as I was writing this book. She told me that the first time that she ever made Toll House cookies was probably when she was 14 or 15. She even changes her recipes from time to time, having moved to the Gold Medal Flour recipe and back to Toll House over the years. Whichever one she uses, she is able to find that all-important step that make her cookies stand out. And, this doesn't stop with cookies, either. Johanna is also better than anyone at making Rice Krispie Treats. While she demurred on the secrets of her cookie making, she confidently told me that the secret to Rice Krispie Treats is to make 1 and 1/2 batches and to heat the marshmallows until they have barely melted. Cooking an amount less than that, or overheating the marshmallows, both result in Rice Krispie Treats that are too hard and difficult to eat. Her real trick may be that she has made one of these recipes at least once a week since she entered college. You can learn a lot about perfecting food when you actually put in the effort to perfecting it. And, now that she has children of her own, Johanna gets to share her magic formulas with her daughters.

While "feel" plays a big role in the quality of your cookies, baking them is an activity that is primed for a little foray into science. Deborah Blum, Pulitzer Prize winning science journalist and publisher of Undark[3] – a magazine for science journalism, had the same problem with her grandmother that I have with my sister; they just "use" the Toll House recipe, but their cookies taste completely different.[4] Inspired by the baking chemistry guru, Shirley Corriher,[5] Blum took to systematically recreating her grandmother's cookies. There are little things that she finds out amid her experiments. Letting the dough rest allows the water to evenly hydrate all of the ingredients, making the dough easier to work with. She also plays with the brown sugar. This is one of the trickiest ingredients for replicating any recipe and really highlights the failure of measuring spoons and cups for portioning out solid ingredients. Brown sugar is loose. It doesn't settle like granulated sugar. The same person could measure brown sugar with the same cups and get a different amount each time. There are two ways around this brown sugar conundrum. Either you measure by mass, which will always give you the right amount. Or, you develop a feel for what measurements you need for a recipe. In the end, Blum writes that the brown sugar is what brought it all back to her. Closing her piece, she muses, "I'm not sure I've discovered either the science or the art – and it's both, as any cook knows – of a perfect chocolate chip cookie."

I already wrote about how my kitchen chemistry classes experiment with cookie recipes (see the Bread chapter). Bethany Brookshire has taken cookie science to a whole new level.[6] She got the great idea to use cookies to engage young students with science. Engagement is kind of an important word here. Some who communicate with science just want to entertain you. Others want you to be actively involved in important or pertinent research. Brookshire's posts were aimed at getting student scientists involved. She wrote about developing hypotheses and ethically testing taste-testers and running statistics, among others. The point of all of her work was not just to show students what research is like; the point of her work was to empower. And, for complex problems like, "How to make the world's best cookie," the best answer will never come from a single person; the best answer will come when lots of people are working toward a solution. This is as true for solar energy and cancer therapies as it is for making chocolate chip cookies.

Meg Hourihan, a tech entrepreneur, took this wisdom-of-the-masses point of view to an extreme. Her recipe is featured in Jeff Potter's delightful book, "Cooking for Geeks."[7] Hourihan collected nearly 50 chocolate chip cookie recipes and found the average amount of all ingredients to make her own. There are lots of normally nonsensical amounts that she calls for: 1/3 of an egg yolk, 2.51 grams of water, 87.9 grams of unsalted butter at room temperature, 29.9 grams of cold unsalted butter, and 58 grams of melted unsalted butter.

Some people take their cookies (and cookie science) very seriously!

Cookies are nice and all, but there is an "everyday" feel about them. Not that I have them every day. But cookies aren't special. Well, that's not true either. Cookies just aren't as special as pie.

Pie lords over all other desserts and is most regal when the crust outshines the filling.

In the pie *versus* cake debate that some people seem to have, the margin of victory isn't even close. Pie wins in a landslide.

I have had a rhubarb pie for my birthday for as long as I can remember. Other than the time that my parents made me an R2-D2 cake, pie has been the only permanent dish on the menu.

There are a million reasons why I love pie so much. But the most important reason of all is the crust. It is flaky and delicate while also being strong enough to stand up to any filling. Cream pies. Fruit pies. Meat pies. Egg pies. A good pastry shell is the only food durable enough to support this diversity while also elevating whatever it contains.

The science of a good pie crust or pastry shell is really interesting, and I'd like to work our way there by starting off at two different extremes – bread dough and phyllo dough – before moving back to pies.

EXTREME 1: BREAD

Bread dough is soft and firm and uniform. Held together by gluten, bread has a consistent consistency that is only interrupted by the exhalations of the baker's yeast. The flour doesn't just turn into an even dough just by adding water. Kneading is needed to make sure that all of those gluten proteins stretch out and line up. When I was a kid, I remember the fund raising campaign that led to "Hands Across America" that was supposed to obtain donations to aid in hunger relief in Africa. The idea was to have people linked together, hand-in-hand, in a show of solidarity that spanned as much of the continental United States as possible. Gluten development is a lot like "Hands Across America." Proteins, much like people, needed to leave their homes and move around until they could start joining onto the chain. In kneading, our physical motions push the proteins around until they can find one another.

There are ways that we can weaken this network. The first is by using lower amounts of protein. Pastry flour and bread flour are both made from soft wheat, which is grown during the winter and has more starch and less protein per grain. The difference between all purpose flour and low protein flour is apparent in the difference between bread and cake. They have a similar structure, but the cake isn't nearly as tough. But, the gluten network within the cake still runs throughout, producing a sponge-like network. We also know that gluten formation is hindered by the presence of oil. When making a roux or flour-based gravy, the flour is coated with oil to ensure that the proteins stay in place on the starch.

A bread dough is roughly 1.7 parts flour for every 1 part water. A cake batter might contain equal parts flour and fat and sugar along with some egg. But, it is evident that uniform and thorough mixing, even when using low protein flour instead of all purpose, and when using oil instead of water, will lead to a consistent crumb.

EXTREME 2: PHYLLO

In the "Joy of Cooking," the directions for making *phyllo* dough basically consist of going to the grocery store and buying premade frozen *phyllo*.[8] The authors write, "it is an arduous and tricky process that yields results no better than what is commercially available frozen in most grocery stores..." Typical recipes for this dough (4 parts flour per 1 part water with a bit of oil) are similar to cracker recipes (3 parts flour per 1 part water with a bit of oil); both are heavy on the flour. This shouldn't come as too much of a surprise, as the dough that supports baklava and spanakopita is often described as being cracker-thin and crispy.

When making and using *phyllo*, the dough is rolled out as thin as possible before receiving a coating of olive oil. Successive layers of dough (or other ingredients) are added to the top of the first layer, which is there to prevent any gluten formation between these layers.

Dishes made with *phyllo* are related to other recipes where thin layers of dough are directly placed on top of one another, separated only by some sort of oil. Another prime example of this is the strudel of German cuisine. Strudels are made by rolling dough as thin as possible before brushing it with melted butter and rolling it over the apple and nut and brown sugar ingredients. Both of these sets of recipes result in wafer thin layers of dough that remain crispy. The layers of dough all span the full dimensions of the dish that is being prepared.

Similar to this, but fundamentally different, is puff pastry. Whereas with the layered dough recipes use a coating of oil for every piece of dough, puff pastries create their own layers through a transformation of geometry. To start preparing a puff pastry, lots of butter needs to be worked into a thin and wide rectangle. The butter needs to be soft enough to spread like this, but cool enough so that none of it runs away as a liquid. This huge piece of butter is placed between two layers of dough, which are pinched closed to seal off the butter. This composite dough is folded into thirds, like a letter about to be placed into an envelope creating, in effect, 3 layers of fat and 4 layers of dough. The folding is followed by the use of a rolling pin to flatten the dough back to its original dimensions. Of critical importance to this process are, keeping the butter warm enough so that it spreads and folds just like the dough, while cool enough so that it does not turn into a liquid and incorporate into the dough. It is important that these be kept separate. Another folding and flattening makes 9 layers of fat and then 81 and then 243 and then 729, each thinner than the one before it and each surrounded by an ever thinning layer of dough. America's Test Kitchen has a wonderful explanation of this process, along with tricks for making it work properly, in their book, "The Science of Good Cooking."[9]

This process is called lamination and is used for all sorts of desserts. Napoleons, bouchées, and turnovers are examples where the lamination leads to thousands of layers of crispy, baked dough surrounded by layers of fat. Siblings of these crispy doughs are croissants and Danish laminated pastries. The reason why croissants and Danishes are soft is that milk and eggs, both of which contain a number of emulsifiers, are used instead

Laminated Dough

In bread dough, gluten forms throughout

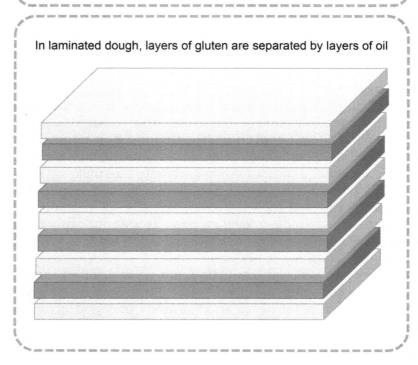

In laminated dough, layers of gluten are separated by layers of oil

of water. Also in this group of pastries are (American) biscuits and scones, which aren't as intensely laminated as croissants but benefit from the folding process nonetheless.

As a research chemist, I am in awe of this process. That bakers could create thousands upon thousands of repeating layers of dough and fat through a seemingly simple folding step is just amazing. They understood that the temperature of the butter was critical to this process. [Butter at 16 °C (60 °F) will crack under pressure and is too cold. Butter at 20 °C (68 °F) will bend and fold as required for the lamination process. At above 24 °C (75 °F), the butter will start to incorporate into the dough.] They also understood that all of the effort that goes into flattening the dough after folding, would necessarily increase the temperature of the butter and so folding should only proceed after the butter has had a chance to cool.

While we have been making 2-dimensional layers in the kitchen for years, we have only recently begun to really craft 2D materials in the research lab. Graphene, a 2D layer of carbon atoms, was isolated from graphite (the stuff that pencil leads, quite the misleading name, are made out of) in 2004. This research spawned thousands of other experiments and eventually won its discoverers the Nobel Prize in Physics in 2010.[10] Just as the thinning layers of dough are different from their origins, so too is graphene different from graphite. Not ones to rest on our laurels, we chemists have been after other 2D chemicals since that time and have even developed ways to synthesize new flat materials from scratch.

PIE

But what about pie? How do we get that consistency? We don't want bread or cake. We certainly don't want perfectly laminated dough. We want something in between. We want flakiness. We don't want parallel layers that run the length of the dough. We want little pieces of randomly oriented laminations. We want layers that start and stop and then start again. The dough needs to come together as a whole, but must be disrupted by pockets of fat. You need to develop some gluten but know when to stop.

And how do we get there, to this state of flakiness? In some ways, the ordered arrangement of layers that we get in lamination makes good sense. We can intuitively understand how that comes to be. The flakiness in good pie crust comes about in a much more random and haphazard way and doesn't really lend itself as easily to logic.

Again, we have the wisdom of ages to help us find our way. The "Joy of Cooking" has an excellent pie crust recipe (using 4 parts butter and 1 part shortening),[8] as does Julia Child (using 3 parts butter and 1 part shortening),[1] as does America's Test Kitchen (using 3 parts butter and 2 parts shortening and some vodka to help prevent gluten formation).[9]

But, this is my book, and I'm going to write about my wife's pie crust, which tops them all.

The first real holiday I spent with Erika was during Thanksgiving the year we were engaged (2001). Both of us were the first from either of our families to have such a serious relationship and my absence disrupted my family's holiday as much as my presence changed that of her family. Both of us were from Ohio, but during that particular Thanksgiving, we travelled to New Jersey to celebrate with Erika's maternal grandmother and her extended family.

That Wednesday night everyone was in a bit of a rush. We were going out for a dinner and so there was the usual struggle to find space in the bathroom to make sure you looked presentable. But, what I remember most was Erika, her mother, her aunt, and her grandmother in their nice clothes, calmly prepping pies for the next day. They were discussing the finer points of peeling and slicing apples. And, later, they prepped the dough before wrapping it and throwing it in the fridge until they were ready to bake the pies.

Both of Erika's grandmothers had an enormous influence on how she has developed as a woman. Her paternal grandmother taught her to do crafts and how to play a prank. Her maternal grandmother instilled in her a love of museums and taught her how to make a pie crust.

Erika has modified this passed wisdom in the time we've been together. What I'm sharing now is her most current preparation. Sift 135 grams (1 2/3 of a cup) of flour. Mix with 1 tablespoon of salt. Using a pastry cutter, mix into the dough 75.7 grams (5 1/3 tablespoons) of butter, 55.5 grams (4 1/3 tablespoons) of shortening, and 12.8 grams (1 tablespoon) of rendered bacon fat. Work the fat and the flour quickly until the mixture resembles small pebbles. Make a paste with 50 grams (1/3 of a cup) of flour and 60 mL (1/4 of a cup) of ice-cold water. Add the paste to the flour-fat mixture until the dough just comes together. Wrap in plastic and store in the refrigerator to allow the dough to hydrate.

Thinking again about what is happening when we make a pie crust. The flour wants to form gluten. That is really its purpose. This recipe allows some gluten to form, but contains it to much smaller spaces. The dimensions of that space are defined by the size of the pebbles after mixing the flour and the fat. Much like oil coats individual flour particles while making a roux, the fat coats a small collection of flour in our developing pie crust. Another trick that gets played on your dough is called *fraisage*. With this technique, the pebbles are streaked across a flat surface with the heel of your hand. In doing this, the pebbles are turned into flake-like shapes even before the dough is rolled and formed into a crust. "The Science of Good Cooking" has a wonderful illustration of the layering and strength of dough that has gone through a *fraisage* step.

I am always amazed at the ability of the chefs that came before us to come up with these recipes. As a chemist, I am certain that if a group of scientists were given the chemical formulas for flour and for butter, in a million years they would not be able to come up with a way to transform these ingredients into a pie crust. But, some chef, hundreds of years ago, figured it out on their own.

Erika uses this dough for any number of different dishes, all of which are family favorites: chicken pot pie, dessert pies (rhubarb, which is my favorite, and cherry), and empanadas.

It is a wonderful and slightly savory crust that pairs as well with meats as it complements and contrasts sweet fillings. The most incredible part of it, though, is the staying power of its flakiness. She has made pies, filled with sweet, fruity, liquidy goodness, that retain their crisp firmness days after they were made.

As the alchemists searched for the essence of different materials, they thought that the elements played symmetrical roles within their physical and spiritual worlds. There is no small amount of alchemy in Erika's ability to elevate. Of course her pie crusts are one manifestation of this. But our children, our family, and our friends are constantly benefiting from the good that she sees in us, the good that we would never have known

PIE DOUGH

Coat flour with butter

Streak the dough (*fraisage*)

Pie dough isn't fully laminated, but the presence of small patches of flour separated by a thin layer of butter is what makes it flaky.

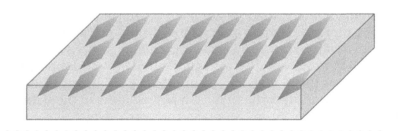

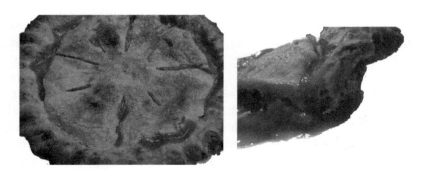

was there. I am constantly being changed by my wife, challenged by her belief in me. She is so quick to see the beauty in other people. I hope that she knows just how much beauty we see in her.

EPILOGUE

I have written, throughout this book, about how chemistry is time and temperature. In some sense, this is true. Time and temperature are the language of chemistry. But, chemistry is really the study of change. How does an ingredient change during cooking? And, how does that change propagate itself through the rest of a recipe.

I am a product of chemistry. I am a product of my surroundings and those with whom I have been surrounded. My experiences, my preferences, and my perception is unshakably linked to my past.

All of the recipes and stories and chemistry in this book are very personal to me.

My biggest goal in writing this book wasn't to share my story; it was to encourage others to see and seek out the chemistry in their lives. Food plays a big role in how I do that.

If you are a non-chemist reading this whose curiosity has been piqued, there are more incredible chemistry stories out there to read. If you are a chemist, the food science that I've talked about here is a mere *amuse-bouche* to the meal that awaits you. Go and seek out more.

But mostly, I hope that you extract a bit of joy from the chemistry in your world, just as I have from mine.

REFERENCES

1. J. Child, L. Bertholle and S. Beck, *Mastering the Art of French Cooking*, Knopf, New York City, 2012.
2. C. Wyman, *The Great American Chocolate Chip Cookie Book: Scrumptious Recipes & Fabled History from Toll House to Cookie Cake Pie*, Countryman Press, Woodstock, Vermont, 2013.
3. *Undark Magazine*, ed. T. Zeller, [document on the Internet]. Undark.org, [cited 2016 July 5], available from: http://undark.org.

4. D. Blum, *So, 268 Chocolate Chip Cookies Later ...* [document on the Internet], Wired, 2011 September 29, [cited 2016 July 5], available from: http://www.wired.com/2011/09/so-268-chocolate-chip-cookies-later/.
5. S. O. Corriher, *BakeWise: The Hows and Whys of Successful Baking with Over 200 Magnificent Recipes*, Scribner, New York City, 2008.
6. B. Brookshire, *Cookie Science* [document on the Internet], Society for Science, 2015, [cited 2016 July 5], available from: https://student.societyforscience.org/collections/cookie-science.
7. J. Potter, *Cooking for Geeks*, O'Reilly, Sebastopol, California, 2010.
8. I. Bombauer, M. R. Becker and E. Becker, *The New All Purpose: Joy of Cooking*, Scribner, New York City, 1997.
9. America's Test Kitchen, *The Science of Good Cooking*, America's Test Kitchen, Brookline, Massachusetts, 2012.
10. The Royal Swedish Academy of Sciences, *Press Release: The Nobel Prize in Physics 2010* [document on the Internet], NobelPrize.org, 2010 October 5, [cited 2016 July 5], available from: http://www.nobelprize.org/nobel_prizes/physics/laureates/2010/press.html.

Subject Index

Note: Page references in *italics* represent diagrams.

acidic solutions 67
acidity 70, 75, 92, 112, 142
acids 64, *65*, 66, 108, 109
    in cheese 106
    in coffee 19
    cooking with 107
    in pectin 92, 93
    in tomatoes 178
    in vinaigrette 135
    see also bases
ACS Reactions (video) 23–24
actin 187, 189, *191*, 200, 202, 203–204
acylpyridines 35
adjuncts 245
Adrià, Ferran 94, 196, 214
Ahern, Daniel 131
Ahern, Shauna James 131
aioli 145
air bubbles 57, 267, 284, 285, 286
alchemists 255, 292, 303
Alchemy of Air, The (Hager) 82
alcohol 236, 240, 243, 246, 248, 265
    see also beer; cocktails; wine
aldehydes 32

ales 245–246
    amber 245
    brown 245–246
    pale 245, 251, 257
alkaloids 8, 9
alkylpyridines 35
Alzheimer's disease 129, 227
American Chemical Society (ACS) 23, 26, 179
America's Test Kitchen 163, 173, 299, 302
    pizza sauce recipe 170
    smoking methods 207
amine group 32
amines 36, 87
amino acids 32, 34, 46, 48, 152, 194
    hydrophilic (water loving) 48
    hydrophobic (water hating) 48, 49, 51
    and Maillard reaction 35
    negatively charged 55, 56, 105, 183
    positively charged 56
    structure of *13*, *47*
    see also proteins
ammonia 67, 82

amylase 241, 242, 243, 248, 250
Andrés, José 44, 196
Andrews, David 21
annatto 100
anthocyanidins 226
anthrax spores 224
antibiotics 28
aquafaba 284
arabica (*Coffea arabica*) 9
arabinose 10
arginine 183, 269
Arnold, Dave 259, 267, 268
aroma molecules 24, 35, 164, 250, 262, 265
aromas 4, 35, 37, 261
artificial fertilizer 71, 82
aspartate 105, 183
aspartic acid 10
Athenaeum 197–198

bacon 23–41, 178
    aroma of 23, 24, 37, 39
    flavor of 29
    frying 29, 30, 37
    Maillard reaction and 30–36
    preparation of 29
bagels 157
baking powder 64, 66, 74
baking soda 64, *65*, 66, 67, 74, 77
baking steel 173
baking stone 173
Barefoot Countessa cookbooks 176, 274
Barham, Peter 189
baristas 6
barley 240–245, 246, 250
    grain 241–242
    and malting *244*

    six-row 241
    two-row 241
Barth, Roger 250
basalt 73
base malt 243, 245, 248
bases 64, *65*, 66
    see also acids
basic solutions 67
Béchamel sauce 114
beef 194, 198–200
beer 235–253
    alcohol volume 255
    brewing 237–240
    flavor 250–251
    see also alcohol; cocktails
bees 9, 225
beets 229–230
Bernbach, Adam 259
betalains 230
betanin 230
bitterness 63, 64, 181, 261
    in beer 251
    in tonic water 254
Bittman, Mark 130, 163
Blum, Deborah 4, 296
Blumenthal, Heston 193, 194, 196
bon appétit magazine 44
bones 200, 211
Borel, Brooke 28
Bosch, Carl 82
botanicals 260, 261, 267, 273
    to flavor vodka 256, 257
    in gin 257, 265
bread 118–134, 284
    bagels 157
    gluten-free 130–131
    no-knead recipe 163–164, 166
    rise time 164

Subject Index

bread dough 127–130, 166, 297–298
   see also phyllo dough
Brenner, Michael 145
brewing
   alcohol content and 250
   all-grain 247
   extract 246
   fermentation 252–253
brine 281
brining 158, 208
brominated vegetable oils 147, 148
Brookshire, Bethany 296
brown sugar 124, 296
   see also sugars
Brunning, Andy 24
buckminsterfullerene 216–217
butanol 141
   see also alcohol
butter 74–75, 101, 143, 193, 299, 301
buttermilk 64, 66, 74, 75
butternut squash 223

Cable, Morgan 224–226
cacao nibs 101
cadaverine 36
caffeine 8–9, 14
cake batter 298
calcium 94, 95, 105, 111, 112, 214
calcium carbonate 70, 73
calcium chloride 168–169, 280
camphene 262, 264, 265
camphor 262
canning 168, 169
caramel sauce 122–123
caramelization 12, 14

carbon dioxide 64, 68, 69, 70–74, 77, 130
   in bread dough 285
   in drinks 259
   in pancakes 67
   photosynthesis and 221
   removal from the atmosphere 73
carbonation 259
carbonic acid *65*, 67–70
carboxylates 87
carotenoids 130, 223, 224, 228
carrageenan 115, 116, 288
Carroll, Ricki 106, 109
cartilage 200
casein proteins 103–106, 107, 109–110, 111, 285
   loose and firm networks *111*
   melting skeleton of 112
   see also milk proteins
catalysts 72, 73
   see also enzymes
Catching Fire (Wrangham) 29
cayenne pepper 115
celiac disease 129, 131
cellulose 10, 87, 207, 269
centrifuges 213
ceviche 107
Chartreuse 266–267
cheese
   acidity of 112
   and charges *111*
   cheddar 102, 111, 113, 114
   Colby 111
   factors that affect meltiness of 112
   Gruyère 102, 114
   macaroni and 100–117

cheese (*continued*)
    melting of 111–112, *113*, 171
    moisture content of 112
    mozzarella 112, 158, 170, 171
    Parmesan 171
    pepper jack 115
    on pizza 170–171
    sauce 101, 102, 111–114
cheese culture 101
cheese curds 106, 109, 112
cheese making 106–107
chemical additives 147
Chemical and Engineering News 7
chemical bonding 151, 183
chemical change 34
chemical interactions 183
chemical reactions 68, 184, 248
chemical stabilizers 288
chemical structures 31
Cheruzel, Lionel 293
chia seeds 132
chicken 42, 179–180, 226
    see also egg whites; egg yolks; eggs
chickpeas 284
Child, Julia
    mousse recipe 293–294
    pie crust recipe 302
chloride ions 55
chlorogenic acid *13*, 14, 18–19
chlorophyll 219, *220*, 221, 224
chocolate 101
chocolate chip cookies 31, 184, 294–297
chocolate malt 246
chocolate mousse 292–294

churning 283–286
chymosin 110
cinchona tree 257
citric acid 106
cocktails 254–273
    the Aviation 267, 269
    foaming 267
    the Last Word 267
    Manhattan 272
    mojitos 272–273
    non-foaming 267
    Penicillin 272
    shaking 267
    stirring 268
    see also alcohol; beer; Chartreuse; gin, and tonic
cocoa butter 101
cocoa powder 101
coconut cream 290
coconut ice cream 290
coconut milk 124, 290
coconut oil 123
Coffea plant 8–9
coffee 3–22, 143, 157
    acids in 14, 16, 19
    aroma 9
    berry harvesting 9–10
    Black Ivory Coffee 11
    brewing 15
    flavor creation 6–15
    flavor extraction 15–21
    grounds 20
    mucilage 10
    Pour-Over technique 21
    preparation of 8
coffee beans
    Bali Kintamani 12
    Ethiopian 8
    fermentation of 10, 11

fresh 12
green 8, 15
Guatemala 8
Papua New Guinea 11
roasted 12, *13*
collagen 200–203, 205–206, 211
colloids 91
Colonna-Dashwood, Maxwell 157
color in cooking 213–231
    Cable's views on 224–226
    Dempsey's views on 221–224
    Keller's views on 227–228
Columbus, Christopher 167
commodity chemicals 71, 72
Compass Coffee (shop) 6–7
Complete Joy of Homebrewing, The (Papazian) 238
condensed milk 103
    see also milk
cone cells 225
confidence 119
continuous phase 142, 143, 144, 145, 151, 152, 153
cookbooks 131, 167, 176, 177, 181
cookie recipes 124, 184, 296, 297
cooking chemistry *see* kitchen chemistry
cooking food 29–30
    benefits of 37
Cooking for Geeks (Potter) 297
Cook's Illustrated 96
corals 70, *71*
coriander, strawberries and 269, 271
cornstarch 75, 85, 88, *97*, 116

adding water to 126–127
uses of 88
Corriher, Shirley 296
Costanzi, Stefano 262
county fairs 26–27
cows 110
cranberry jelly 92
cream 58
crème brûlée 58
Crème de violette 267, 268, 269
crock-pots 199
croissants 299
crystal malts 243, 246
Cup4Cup 131
curcumin 228, *229*
curds *see* cheese curds
custards 287–288
cyanin *226*, 226

Danish pastries 299
Dayton, Ohio 61
deconstructing foods 101, 113
Dempsey, Jullian 221–224
digestion 30, 38
diseases, aging-related 129
dispersed phase 142, 143, 144, 151, 152
distillation 255–256, 277–278
DNA 46
dogs 225
donuts 61–62
dough *see* bread dough; *phyllo* dough
Drunken Botanist, The (Stewart) 265
dry aging, of meat 194–195
Dutch oven 164
dyes 100
    see also pigments
dysprosium 224

egg proteins 55, 56, 58, 287, 294
egg whites 46, 57, 284, 285, 294
egg yolks 51, 52–53, 54, 143, 294
    in icecream 287
    in macaroni and cheese 116
    in mayonnaise 152
eggs 42–59, 103, 116–117, 288
    $6X\,°C$ egg 51
    cooked *52*
    cooking food containing 45–49
    cooking with acid 107
    effect of heating on 51
    overcooked *52*, 54–55
    raw *52*
    research on 49–54
    runny 54
    scrambled 42, 43, 45, 54, 55, 57–58
    unsalted *56*
    whipped 57
electric charge 105
electrolytes 146–147
electrons 138–140, 218–219
electrostatic interactions 183
emulsifiers 53, 115, 148, *149*, 285
    amphiphilic 148–150, *149*, 151, 152, 287, 294
    hydrocolloid *149*
emulsions 16, 142–143, 144, 146, *149*, 285
energy 82
    in molecules 81
    storage of 85
enzymes 110, 194
    see also amylase; catalysts
equilibrium 32, 34–36, 68–69, *69*

espresso 6, 16
Essentials of Classic Italian Cooking (Hazan) 199
Essentials of Italian Cuisine, The (Hazan) 167, 289
Estabrook, Barry 28
esters 68–69
ethanol 136, 140, 243, 255, 256
europium 224
evaporation 255–256
evolution, food safety and 36–38

failure 121–122
Falkowitz, Max 276, 279, 288
farmers 27, 28
farming 26, 28
fast food restaurants 215
fermentation 240, 252
    coffee 10, 11, 12, 20
    yeast 226, 246
fertilizers 71, 82
filets 200
fish 208
flans 58
flavor pairings 270
flavor(s) *33*, 261, 270
    biochemistry *263*
    development of *165*
flaxseed meal 132
flour 126, 127
fMRI 214
foams 95, 132, 267, 283–285, *286*
food additives 96
Food Biophysics 51
food ethicists 27, 28
Food Lab, The (Lopez-Alt) 116
food pairings 269–270
food safety, and evolution 36–38

Subject Index

food science 4, 206
food scientists 27, 28, 146
fossil fuels 84
fraisage 303
fresh foods 146
fructose 10, 31
fruit caviars 94, 95, 214
fruits 88–89, 91–92, 228
furanones 35
furans 35

G protein-coupled receptors 261–262
galactose 10
galacturonic acid 89, 92
Garten, Ina 274
gases 144
gelatin 91, 202
gelatinization 284
gelato 289
gin 254, 255, 256–257, 264–265
    and tonic 4, 254, 255, 259–260, 262
gliadin 125, 126, 127, 132
global warming 70, 72
glucose 10, 31–32, 83, 87, 89
    energy storage and 85
    respiration and 84
    structure of *86*
glutamate 105, 183, 194
glutamic acid 10
gluten 85, 125–126, 127, 297, 303
    arrangement of protein fibers in 129
    and kneading *128*
    in pizza dough 166
    roles in bread 130
gluten-free bread 130–131

Gluten-Free Flour Power (Kamozawa and Talbot) 132
Gluten-Free Girl (blog) 131
glutenin 125, 126, 127, 130, 132, 133
glyceryl trioleate 140
grains 35
graphene 301
graphite 301
Gray, Harry 217–218
Great Barrier Reef 70
guacamole 177
guar gum 132, 290
gypsum 282

Haber–Bosch process 82
Haber, Fritz 82
Haft, Michael 7
Harvard University 145
Hazan, Marcella 167, 168, 199, 289
heat 185
    see also temperature
heat exchangers 279, 281
heat transfer 190
heating rates 189
heme group 204
hemoglobin 204
Hendon, Christopher 157
(Z)-3-hexanal 269, 271
hexanol 141
hogs 26, 27
homogenization 285
hops 250–251, *252*
hot chocolate 102
Hourihan, Meg 297
humidity 166
humulone 251
Humulus lupulus 250
Huntington's disease 129

hydrochloric acid 66
hydrocolloids 90–91, 95, 96, 98, 131
    as emulsifier 148
    tomato sauce, thickening by 169
hydrogels 52, 54, 56, 58
hydrogen ion *see* protons
hydrophilic (water loving)
    molecules 17, 53, 117, *137*, 146
        in coffee 19
        definition of 136
        examples of 140–141
        in sports beverages 147
        in starch 127
hydrophobic (water hating)
    molecules 17, 18, 53, 127, *139*, 140
        amino acid 48
        in coffee 19
        examples of 141
        in food 146
hydrothermal vents 82
hydroxide ion 66

ice 90, 277, 280, 281
ice cream 58, 274–291
    base malt 287–288
    churning 283–286
    coconut 290
    fat in 285
    foam 285
    freezing 276–282
    gelato and 289
    liquid nitrogen 288
    overrun 286, 289
    Philadelphia-style 288
    recipes 275–276
    standard 288
    tips for making 279
    vanilla 274
ice cream makers 274
    heat exchangers 281
ice crystals 277, 279, 281–282, 283, 287
Ideas in Food (Kamozawa and Talbot) 98, 115, 131
"Imposter Syndrome" 119
ionic compounds 138
iridescence 16, 17
isohumulone 251
Italians 167

jams 89, 91, 95
Jarrold, Martin 119
Jell-o 91
jelly 81–99
Jerusalem: A Cookbook (Ottolenghi) 222
journalists 28
Joy of Cooking 298, 302
Jubany, Nandu 145
juniper berries 256

Kamozawa, Aki 98, 115, 131
Keller, Gretchen 216, 227–228
Keller, Thomas 131
Khymos (blog) 96, 98, 269
kinetics 185
Kitchen as Laboratory, The (César) 50, 51
kitchen chemistry 238
kneading *128*, 129–130, 164, 297–298
Krispy Kreme 61
Kurti, Nicholas 94

lactic acid *65*, 66, 67, 75
Lactobacillus bacteria 75
Lactococcus bacteria 75

Lahey, Jim 163, 166
laminated dough *300*
lamination 299
Lang, Adam Perry 193
Lebovitz, David 275, 289, 290
lecithin 117, 152, 287
lemon juice 77, 93
Lersch, Martin 95–96, 98, 269
Leuconostoc bacteria 75
Libbrecht, Kenneth 281
ligaments 200
lignin 207
liqueurs 266, 267–268, 269
Liquid Intelligence (Arnold) 259, 267, 268
liquid nitrogen 277–278
liquids 144
liquors 255
Lopez-Alt, Kenji 163, 173, 187
    dry-aging beef 194–195
    experiments with vinaigrettes 152–153
    macaroni and cheese recipe 116–117
lysine 35, 183

macaroni and cheese 100–117
    melting cheese 111–117
    recipes 113–116
    sauce 101–102
    see also cheese; milk
McDonalds 215
McGee, Harold 55, 110, 129, 167
    method of cooking steak 192
    vinaigrette recipe 136, 142, 152
McKenna, Maryn 28
Madison, Deborah 63, 74, 75
magnesium chloride 280

Maillard, Louis-Camille 31
Maillard reaction 30–36, 37, 38, 39, 130, 151
    in bacon 26
    in coffee beans 12–14
    in cooking steak 190–192
    for grain products 35
    mashing and 250
    roasted malts and 243
    smells produced by 30–31
    smoking and 210
malaria 257
malt 246
malt extracts 246
malting 242–243, *244*
mangoes 177
mantis shrimp 225
maple syrup 76
maraschino liqueur 267–268
mashing 247, *249*
mayonnaise 143, 152
meals, mind-changing 196
measuring spoons 186
meat 35, 176–195, 196–212
    chewiness/tenderness of 183
    cooked 35, 37, 38
    cooking 187, *188, 191*
    dry aging of 194–195
    red color of 204–205
    salting and *182*
methanol 140
methoxy ester group 89
milk 58, 101, 102–103, *104*, 143
    emulsifiers 115
    evaporated 103, 115, 116, 288
    film 102
    phosphate groups 108
    preservation of 110
milk powder 133

milk proteins 103
    see also casein proteins
milkfat 101, 106, 112
Miller, Alex 71, 72, 73, 210
minerals 282
modernist cuisine *see* molecular gastronomy
Modernist Cuisine (Myhrvold) 6, 94, 206, 208
molecular gastronomy 50, 94–95, 96, 132, 213
molecular motion 144
monosaccharides 85, *86*
Morrison, Jessica 7
mousse 292–294
MRI 214
mucilage 132, 153
muscles 194
mustard 153
My Paris Kitchen (Lebovitz) 275
Myhrvold, Nathan 6, 94, 206, 208
myoglobin 190, 204–205, *205*, 209
myosin 187, 189, *191*, 200, 203, 204

nature *versus* nurture debate 46
NBC 38, 39
negative charges 55, 56, 106, 108, 109, 168
    in jam making 93, 94
    in meat 183
    in proteins 105
    see also electrons
New York Times, The 163, 164
nitric oxide 209
nitrogen 82
non-Newtonian materials 126

oats 243
odor molecules 264
odor receptor proteins 264
oils 117, 140, 141–142, 146
    and gluten formation 127
    mixing water and 135
olive oil 135, 142
omelets 44–45
omniphobic materials 140
On Food and Cooking (McGee) 55, 110, 129, 136, 142, 167
Oobleck 126–127
Orangette (blog by Wizenberg) 75
Ottolenghi, Yotam 222
oven temperature 174, 186
overrun 286
oxazoles 35
oxygen 130, 204

pancakes 60–78
    recipes 63–64, 76–77
Papazian, Charlie 238–239
paprika 100
Parkinson's disease 129
pasta, with braised bacon and roasted tomato sauce 178
Pastan, Ira 173
Pastan, Linda 173
Pastan, Peter 173–174
pathogens 205
pectin 10, 88–89, 91, 92, 94, 95
    high-methoxyl 92, 93
    low-methoxyl 92, 93, 94
pectinases 10–11
pentanol 141
Penthouse Executive Club 193
Perfect Meal: The multisensory science of food and dining, The (Spence and Piqueras-Fiszman) 214

Subject Index 317

periodic table 218
Peruvians 257
pH scale *65*, 67
phosphate groups 105, 106, 108–109
photosynthesis 81, 83, 84, 85, 221, 241
phyllo dough 298–301
    see also bread dough
pies 297, 301–305
    flakiness of 302
    recipes 302–303
pig feed 29
Pig Tales (Estabrook) 28
pigments 230
    see also dyes
pigs 26–30
    see also bacon
Piqueras-Fiszman, Betina 214
pizza 157–175
    baking *172*
    California 159
    chains 161–162
    Chicago style 159, 161
    cooking time 172
    crust 163–167
    dough 158, *165*, 166
    English muffin 158
    Neapolitan 163, 165, 173
    New York style 159, 163, 165, 166
    ovens 165, 171
    St. Louis 159
    sauce 167–170
    slices 160
    soggy 172
    toppings 170–171
    USA regional variations in 159
pizza napoletana 159
plaques 227

plating food 216
polarity 138
polymers 87
polysaccharides *86*, 89, 91, 115, 126, 247–248
popcorn 88, 283–284, 285
popovers 76
pork
    salting 29
    shoulder 208
    smoking 29
porters 246
positive charges 48, 56, 93, 168, 183
pot roasts 167, 199, 200, 202–203
potato starch 85
Potter, Jeff 297
pre-ferments 164, 165, 166
precision 50
processing food 146
professional chefs 27
proline 35
propanol 141
propylene glycol 280, 281
proteases 194
proteins 36, 46, 48, *108*, 227
    architecture *47*
    in barley 240
    effects of temperature on 51, 103
    extended 105
    shapes of *47*, 48
    unfolding of 187–189
    see also amino acids
protons 66, 67, 68, 108, 109
Puck, Wolfgang 44
puddings 88, 89, 90, 91
puff pastry 299
putrescine 36
pyrazines 35

pyridine 36–37, 39
pyrolysis 207
pyrroles 35

quantum mechanics, rules of 218
quinine 257, 261, 265

Ramsay, Gordon 45, 54, 55, 57
Rasmussen, Patricia 281
Ratatouille (film) 5, 124, 197, 260–261
Ratio (Ruhlman) 133
Ratner, Mark 120
reactions, analysis of 66
recipes
    following 118
        reverse engineering of 122
    substitutions in 124
reconstructing foods 101
reducing groups 32
rennet 110–111
research chemistry 238
respiration *84*, 84
retrosynthesis 287
reverse-searing 192
Rice Krispie Treats 295
rise time, of dough 130, 164
roast beef 198, 199, 200
roast chicken 179–180
roasted malts 243
roasting
    coffee beans 7–8, 9, 12, 14
    meat 205
Robusta (*Coffea canephora*) 9
rock salt 280
Rodgers, Judy 177, 180, 184
Ross, Sam 272
rotary evaporators 213

roux 113–114, 127
Ruhlman, Michael 133, 166
rye 243

Sadowitz, Lena 21
Sagan, Carl 83
salmon, smoking 208
salt 55–56, 58, 136, 183, 280
salt water 280
salting food 180, 181
    meat *182*, 183–184
    pork 29
salty taste 181, 261
samarium 224
San Francisco 178–179
savory (umami) taste 181, 261
Science (journal) 73
Science of Cooking, The (Barham) 189
Science of Good Cooking, The 299, 303
scotch 272
seaweeds 115
senses, involved in eating 215
sensory sciences 214
Serious Eats (blog) 276, 279
shortbreads 124, 127
sight 215
smoke ring 209, 210
smokers 207, 208–209, 210, 211
smoking food 206–211
    pork 29
    ribs 209–210
snowflakes 281, 282
soap 150
sodium alginate 94
sodium bicarbonate 64, 74
sodium chloride 136, 140, 280, 282

sodium ions 55
sodium lactate 67
solar energy 221–222
solids 144
solutions 67
sorbet 288–290
Sörensen, Pia 145
sound, and food 215
sour taste 64, 181, 261
sourdough bread 75
sourdough starter 164
sous vide cooking 50, 192–193, 208, 213
Spence, Charles 214
spherification 214
sports beverages 146–147
squash 223, 224
stabilizers 288, 290
starch 85, *86*, 87, 240
steak 186–187, 189, 190, 192–193, 200, 203
Stewart, Amy 265
stocks 177
stouts 246
Strand, Oliver 164
strawberries 269
    and coriander 269, 271
    flavor molecules 270
strudels 299
Suarez, Harrison 7
sucrose 31, *86*, 138, 140, 141
sucrose acetate isobutyrate 148
sugars *13*, 31, 36, 136
    fermentable 247
    in icecream making 287
    in jam making 93
    non-fermentable 247–248
    simple 10

    see also brown sugar; fructose; galactose; glucose; sucrose
sulfur 130
sulfuric acid 217
surfactants 150
sweet taste 261
syrups 136–138

table sugar *see* sucrose
Talbot, H. Alexander 98, 115, 131
tapioca starch 85
taste 180–181
temperature 184, 203, 305
    effects on proteins 51
    oven 174, 186
tendons 200
terbium 224
texture 90
Texture (Martin) 96, 98
thermodynamics 184–185
    second law of 279, 281
thiophenes 35
This, Hervé 94
time 184, 305
Timón, María 24, 26, 38, 39
tinctures 266
Titan 225
Today Show, The 23, 24, 40
tofu 228
tomato sauce 168–170, 178
tomatoes 158, 167, 168–169
tonic water 254, 257, 261
tonics 257
transition states *65*, 68
turmeric 100, 228, *229*
twins, identical 46

vanilla 14
Vega, César 49–54, 57, 279
vegan caramel sauce 123–124
vegetables 88, 216, 218, 228
Vegetarian Cooking for Everyone (Madison) 63
Velveeta 101–102
vinaigrettes 16, 17, 135–154
    how to make 142, 152–153
    recipes 135–136
vinegar 136, 142
viscosity 53
visible spectrum 218, 219
vodka 255, 256, 257

waffles 88
    recipes 75–76
    toppings 76
Warner, Brandon 7–8, 11, 14–15
Washington Post 7
water 49, 67, 89–90, 136, 138, 158
    boiling temperature 255, 256
    freezing 277, 280, 281–282
    and gluten formation 127
    ice 90
    in meat muscle fibers 183
    mixing oils and 16–17, 67, 135, *142*
    in pizza 157

Weber kettle grill 207, 208
Weitz, David 145
wheat 243
wheat starch 85
    see also flour
whey 101, 106
whey milk protein concentrate 101
whey protein concentrate 101
whey proteins 103
whipped cream 76
whisky 271
wine 253
    color of 226
    red and white 215–216
Wizenberg, Molly 75
wort 250, 251
Wrangham, Richard 29–30, 36, 37

xanthan gum 131, 132
Xanthomonas capestris 132

yeast 20, 111, 164, 174, 226, 252, 255
yellow 5 dye 100
yellow 6 dye 100

Zuni Café 179, 183
Zuni Café Cookbook, The (Rodgers) 177–178